Lesley Fairbairn.

Cambridge Elementary Classics

THE STORY OF PALLAS

AENEID VIII, 26–183, 280–369, 454–607;
x, 362–79, 439–509; XI, 24–99,
139–81; XII, 930–52

*With introduction, notes, vocabulary
and illustrations*

COMPANION VOLUMES
THE STORY OF CAMILLA (1956)
THE STORY OF TROJAN AENEAS (1959)

THE STORY OF
PALLAS

❊❊❊❊❊❊❊❊❊❊❊❊❊❊❊❊❊❊❊❊❊❊❊❊❊❊❊

EDITED BY

BERTHA TILLY

M.A., PH.D.

Headmistress of the High School for Girls, Ely

FROM

VERGIL'S *AENEID*, BOOKS VIII, X, XI & XII

CAMBRIDGE
AT THE UNIVERSITY PRESS
1961

PUBLISHED BY
THE SYNDICS OF THE CAMBRIDGE UNIVERSITY PRESS

Bentley House, 200 Euston Road, London, N.W. 1
American Branch: 32 East 57th Street, New York 22, N.Y.
West African Office: P.O. Box 33, Ibadan, Nigeria

CAMBRIDGE UNIVERSITY PRESS
1961

Printed in Great Britain at the University Press, Cambridge
(Brooke Crutchley, University Printer)

CONTENTS

Illustrations	page vi
Preface	vii
Introduction	xi
(i) Vergil's life and works	xi
(ii) The story of the *Aeneid*	xvii
(iii) Pallas	xix
(iv) The hexameter	xxii
(v) Reading and translating Vergil	xxiv
THE TEXT	
Book VIII, lines 26–183, 280–369, 454–607	1
Book X, lines 362–79, 439–509	21
Book XI, lines 24–99, 139–81	25
Book XII, lines 930–52	31
Notes	33
Notes on background	118
Vocabulary	136

The drawing on the cover is adapted from an Etruscan ash chest of terracotta on which is shown a dying huntsman. A shield and helmet have been added in place of the hunting dog in the original.

ILLUSTRATIONS

THE PLATES ARE BOUND TOGETHER BETWEEN
PAGES 6 AND 7

I   (*a*) Tiberinus (*Museo delle Terme, Rome*)
    (*b*) The Tiber and the Aventine Hill

II   The Palatine Hill

III   The Roman Forum from the Capitoline Hill

IV   The *Clivus Capitolinus*

V   Remains of the Temple of Saturn

VI   The site of the *Forum Boarium*

VII   The Capitoline Hill

VIII   The Punishment of Tarpeia (*Unione Fototeca, Rome*)

The map on p. 129 is a simplified version of a map in Gösta Säflund, *Le Mura di Roma Repubblicana*, and is reproduced by permission of the Publication Board of the Swedish Institute in Rome.

PREFACE

I want my readers, in particular those who read this work as a task for an examination, to enjoy the poetry they will study. I have therefore edited this book in a particular way, convinced that the time has come for a more illuminating approach to the reading of Latin texts than has been customary. Provided that the Latin is correctly understood, I do not believe in discussion of grammatical niceties, linguistic rarities, and pieces of erudition which are, after all, a natural outcome of the poet's expression, but in simple explanation and even, in places, translation of more difficult passages. References to Greek literature which are meaningless to one who knows no Greek, and to other Latin writers who do not come within the ken of the average student, have been purposely omitted. To me the background of the poem is far more important; I mean the elementals of the Italian climate, the brilliant, unfailing sunshine, clear air and bright skies which were the poet's life experience, and the natural images of the countryside, such as the ordered cultivation of the land, vineyards, olive groves and carefully planted crops, the sheep with their shepherds and pastures, and, never far away, the sapphire sea of the Mediterranean. All these solaced and delighted Vergil in his imagination and will lead us to a deeper understanding if we are open to them. To these must be added such accounts of Roman customs, religion, art and archaeology as seem necessary for proper

PREFACE

understanding of the text. We do not study Shakespeare, Milton, or any other great English poets as if their poetry contained museum pieces of linguistic rarities, but in the joy of understanding the expression of their urgent thought. In just this way Vergil too can be loved: we should seek him with delight.

In order to reach such a degree of appreciation of Vergil's poetry, the beginner must be guided to a quicker entry into the magic city, for there is much magic there. The notes must assist and direct his steps so that encouraging progress may be made from the first and a desire left for more. Discouraging attempts, for instance, to identify unfamiliar parts of verbs, to recognize agreements which do not quickly make themselves clear; bewilderment at terms of thought too subtle for the immature, and at long sentences of which the construction is abstruse; all these often choke off interest from the first, and cause the study of Latin to become the bugbear which it need not be.

The old annotated texts, such as those of J. Conington and H. Nettleship, A. Sidgwick, and T. E. Page, packed with excellent learning and scholarship, were the masterpieces of a hundred, seventy, or sixty years ago, but were written for the public schoolboy who learned little else than Latin and Greek all his schooldays. These and others, together with the works of Servius, the fourth-century commentator, have been constantly before me, and I hereby express my indebtedness to them. They do not, however, meet the needs of today. If Latin, menaced more and more by the increasing need for scientific and technological studies, is to keep even the restricted place in the

PREFACE

curriculum of the average school which is usually allotted to it, the need for texts with more help and guidance and with less erudition is critical. I even go so far as to claim that such as these could be instrumental in preserving, and even furthering, the study of Latin in the face of modern requirements. At the same time I suggest that the old annotated texts mentioned above should be at hand for the more advanced student.

In this book, which is a companion volume in this series to *The Story of Camilla* and *The Story of Trojan Aeneas*, passages have been assembled to tell the tragic story of Pallas, the son of Evander, the Greek settler on the Palatine Hill before the days of Rome. He is a young, untried, eager victim of war, who dies in earliest manhood for the Rome of the future. It is a story which sounds down the centuries the futility of war. In it too is the sorrowful figure of Evander, poor, infirm, who relinquishes his only son for a great cause. His helplessness and despair, increased by the weakness of old age, make him comparable with the tragic figures of great literature. Although the plots have no affinity, it is suggested that passages in translation from the *Oedipus Colonus* of Sophocles, and from Shakespeare's *Lear*, be read alongside the *Story of Pallas*.

In the notes more help than is usual has been given in view of the scanty amount of time which is usually available for the preparation of a text. The notes which fill in the essential Roman background are for the student working for the Advanced Level of the General Certificate and for the

PREFACE

teacher. The length of the book as a whole (633 lines) makes it suitable to be a set book at this level: it could be set for the Ordinary Level too, with the omission of certain excerpts. The book will also be found useful for Sixth Form reading apart from examination requirements, and first year University students might well find it a welcome supplement to more detailed study of texts.

My gratitude is acknowledged to Mr L. Roberts, Director of the American Academy in Rome, for the privilege of residence on several occasions and the use of the Library; to the Oxford University Press for permission to use the Oxford text of the *Aeneid* (this has been followed exclusively, with the exception that the ending *-es* of the third declension accusative plural has been substituted for the less usual *-is*); to Professor E. T. Salmon for his help concerning the topography of ancient Rome; to Mr Nalder Williams for his kind scrutiny of the manuscript and many helpful comments; to all those who have encouraged the writing of this series, and especially to Mr O. C. Watson, lately of the Cambridge University Press, whose advice has always been most helpful; and to the Syndics of the Press.

BERTHA TILLY

AMERICAN ACADEMY IN ROME
August 1958

INTRODUCTION

(i) *Vergil's life and works*

Vergil, the greatest Roman poet, is numbered among the greatest in any language the world has ever known. He was the interpreter of the most glorious age in the history of Rome, one often called 'The Golden Age', when the emperor Augustus established the Empire and brought peace to a world thirsting for rest after years of civil war. It was during the reign of Augustus, but some fifteen years after Vergil's death, that Jesus Christ was born in Bethlehem in the Roman province of Judaea.

Publius Vergilius Maro was born on 15 October in 70 B.C. at a small village called Andes (usually identified with modern Pietole) near Mantua in north Italy, a district which was then called Cisalpine Gaul. His father, a farmer, possessed his own estate, and here from early childhood he learned to love country life and all the scenes and pursuits that belonged to farming. Grown to boyhood, he was given the best education which could be found in those days. He went to school first at Cremona, a city not far from his home, and from there for a short time at Mediolanum (Milan) where the best schools in Cisalpine Gaul were to be found. After this, as a young man, he received the equivalent of a university education in Rome. Here between 55 and 50 B.C. he practised eloquence and learnt rhetoric (the art of public speaking) from the famous teacher Epidius, and others. In the lecture-room Vergil sat beside sons of the most distinguished families, among whom, it is thought, may have been the young great-nephew of Julius Caesar—younger than Vergil by some seven years, but advanced in wisdom and attainment—Octavius, the future emperor Augustus. The studies he pursued in Rome were clearly intended to prepare him for a public career such as that of the law-courts or politics, but he wearied of the exacting artificial instruction of the rhetoricians. Consequently, in his early twenties he turned to philosophy and sought the teaching of the Epicurean Siro in the academy called 'The Garden' at Naples. Here, in peace, he became familiar with the philosophic thought which colours all his poetry. It is generally thought that another great Roman poet not much older than Vergil studied in this very place, and learned the same way of thought. Titus Lucretius Carus, who expounded the teachings of Epicurus in verse only surpassed by Vergil's own, lived from 95 to 55 B.C. His great work the *De Rerum Natura*, 'Concerning the Nature of Things', published during Vergil's boyhood, exercised a deep influence on his poetry.

When his student days were ended, Vergil returned to his home in the north, and lived there until 42 B.C. For many years before this Italy had been torn by civil wars so that his youth must often have been saddened by rumours and scenes of death and devastation, but that particular year brought a climax of anxiety for his family. The Triumvirs Octavianus (the future emperor Augustus), Marcus Antonius and Lepidus defeated Brutus and Cassius, the assassins of Julius Caesar, at the decisive battle of Philippi. The

INTRODUCTION

Roman world lay at their feet and they promised land to their victorious soldiers. In the settlement which followed, the farm at Andes was seized for occupation, to be parcelled out to veterans of the wars. The family property, however, was saved when Vergil himself gained an introduction through Pollio and Varus, successive governors of Cisalpine Gaul, to Octavian, and through his intervention obtained the restoration of his father's property.

For some time, perhaps during these years, he lived in Rome and had a house on the Esquiline Hill, but being a lover of quiet, and a man of a contemplative disposition, he withdrew early in life from the noise and tumult of Rome to make his home near Naples. Here, dedicating himself to poetry, he spent the rest of his days on his own estate, enjoying seclusion and peace in his Campanian retreat. Always before him was the sapphire blue of the waters of the Bay of Naples, around were the ordered agricultural scenes of the most fertile districts in Italy, and behind the sheltering Apennines; unfailing sunshine and warmth were always his. Through his poetry he soon became one of the brilliant circle of poets and writers who enjoyed the patronage of Maecenas, the minister of Augustus, and the friend of literary men. Vergil also won the favour of Augustus himself. Little is known of his life, and still less of his character except that he was pure, simple, and affectionate: the poet Horace, one of his close friends, calls him *animae dimidium meae*, 'the half of my soul'.

Apart from some minor works written in youth, his first published works were the *Eclogues*, ten short poems

INTRODUCTION

in which are described scenes of country life: these are partly imaginary, and partly pieces alluding to contemporary events such as the loss and restoration of his farm, and mentioning among others his friends Pollio and the soldier-poet Gallus. This kind of poetry, called 'pastoral' because shepherd life is most usually described, originated with the Greek poet Theocritus who lived in Sicily in the third century B.C. It was Vergil who extended this form to present living persons in rustic appearance. There are many instances of 'the pastoral' in English literature, especially in the poetry of Spenser and Milton, and in Elizabethan lyrics. The *Eclogues* were published in 37 B.C. and had an immediate success. Vergil was hailed as a poet of future greatness, judged to have brought a new freedom into Roman poetry. He was a 'modern', and his poems, full of charm and elegance, were a fitting prelude to greater achievements. The *Fourth Eclogue* is called the 'Messianic Eclogue', because there is foretold in it the birth of a child who would bring back to earth the Golden Age. Although it cannot be known for certain which child was meant, a general mistaken belief grew up later that Vergil was writing of the birth of Christ. For this reason he gained the reputation of being a prophet of Christianity, one that lasted through all the Middle Ages.

One of the first aims of Augustus on his accession to power was to revive agriculture—which had declined during the long-drawn-out years of the civil wars—in an endeavour to bring back an era of prosperity to the land. Vergil's next works, the four *Georgics* (poems of country life), were written at the prompting of

Maecenas, reflect his reawakening of interest in the Italian countryside, and are an incentive to remember the great days of the past. Vergil spent the next seven years from 37 to 30 B.C. in composing these poems on farming and its different branches: the First Book deals with agriculture proper, the Second with the cultivation of trees, especially the vine and the olive, the Third with the breeding of cattle and horses, and the Fourth with bee-keeping. All through the poems there sounds the true love of a countryman for the very soil and all the varied scenes and activities of the farmer's life. A sweetness and sympathy for nature which are peculiarly Vergilian prevent them from being merely 'didactic' or instructive. In some passages occur splendid descriptions of the Italian countryside. Vergil took his theme from the Greek poet Hesiod who wrote the *Works and Days*, and was influenced at the same time by the great philosophic poem of Lucretius, but the language and style are unique, and all his own. The *Georgics* are perhaps his most artistic work: the language is of the greatest felicity, and the metre reaches the perfection of the Latin hexameter.

In 31 B.C. Octavianus returned to Rome from Egypt, victorious over all his opponents after the defeat of Antony at Actium; under the title of 'Augustus' he ushered in the new age of Rome and of the Empire, and established peace throughout the world. At his bidding, Vergil began to contemplate the writing of a national epic which would exalt the achievements of Augustus. His plans began to take shape between the years 29 and 27 B.C. This poem, far more ambitious than his previous works, was to link

INTRODUCTION

Troy and the heroic age with the foundation of the City of Rome, and the family to which Julius Caesar and Augustus both belonged. This poem, the *Aeneid*, composed during the last ten years of his life, contains the deeds of a mythical hero, Aeneas, who fought against the Greeks at Troy, and then, as a refugee from the burning city, voyaged to Italy and there became the founder of the Roman race. Aeneas is shown as the ancestor of Augustus and the founder of the *gens Iulia* (the Julian family) to which Augustus' father by adoption, Julius Caesar, belonged. Aeneas is characterized as a man devoted to duty, ready to sacrifice all for the mission which the gods have set before him, and to endure all kinds of hardships and dangers. In him is to be recognized Augustus in a heroic form: the mythical figure seems to make him grander, more than human.

Vergil was strongly influenced by Homer's *Iliad* and *Odyssey*, but he succeeded in composing a poem which is wholly Roman in spirit and outlook. In it is the consciousness of Rome's great destiny to rule the world, and of that peculiar resurgence of patriotic feeling which belongs to the Augustan age. Vergil had been writing the *Aeneid* for ten years when he undertook a journey to Greece and Asia Minor to see the places of which he had written. On the return journey, travelling in the company of Augustus, whom he met at Athens, he fell ill, and died on reaching Brundisium in southern Italy in 19 B.C. The poem was still incomplete, and although he is said to have left orders to his executors for it to be destroyed this was (fortunately) prevented by the intervention of Augustus, who in this way preserved for all time one of the greatest poems in all literature. Vergil was buried near Naples, where he had lived for so many tranquil years, on the road to Puteoli. On his tomb was inscribed the epitaph which he was said to have composed for himself:

> Mantua me genuit, Calabri rapuere, tenet nunc
> Parthenope; cecini pascua, rura, duces.

(ii) *The story of the Aeneid*

The *Aeneid* is made up of twelve books. In imitation of Homer's manner the poem begins at a point far on in the story in such a way that the earlier part of the narrative is told by Aeneas himself. In Book I the Trojans, who have passed a long time in voyaging in the eastern Mediterranean in search of the land promised them by the gods, set sail from Sicily and are driven by a storm to the shores of Africa. There they meet with the queen of Carthage, Dido, who gives a banquet in their honour and asks Aeneas to tell of their adventures. The next books, II and III, are taken up with his story: in Book II he tells of the fall of Troy and of his escape from the burning city with his father Anchises and his son Ascanius; in Book III he tells how they landed at one place after another, hoping each time that they had at last come to their rightful home, until at last they reached Sicily where Anchises had died. In Book IV is told the love of Dido for Aeneas, how he stayed in Carthage for a year, but then, reminded of his destiny, set sail. Dido in despair slays herself on her own funeral pyre. In Book V the

Trojans reach Sicily again and celebrate funeral games in honour of Anchises. Book VI brings them to Cumae, where Aeneas descends to the underworld to visit the shade of Anchises. He finds him in 'the fields of the blessed' and is shown a pageant of Roman heroes.

The first six books thus tell of Aeneas's long wanderings in search of the promised land of Italy. The last six tell the story of the Trojans in Italy and of the war which was fought between them and the inhabitants of Latium. It ended only with Aeneas's victory in single combat over the enemy's champion, the Rutulian Turnus. In Book VII the Trojans reach Italian soil at last and land at the mouth of the Tiber. There they build a camp and then reconnoitre the country. At the same time the native king, Latinus, is told by an oracle that his daughter Lavinia is not to marry Turnus, to whom she is already betrothed, but will be the bride of a foreign prince. He later recognizes Aeneas as her destined husband. Then the war between Latins and Trojans breaks out and there follows the 'gathering of the clans'. The native tribes muster from every direction to the plain of Latium to withstand the Trojan invasion. They come in picturesque and glittering array, making a splendid pageant of armed warriors. At their rear comes the Amazon Camilla, a fine athletic figure, who rouses admiration in all who behold her. Book VIII describes how Aeneas goes to seek the help of King Evander, who lives on one of the seven hills of future Rome. He arrives in time to take part in the yearly sacrifice to Hercules at the *ara maxima* on the Tiber's bank. Evander takes him to his home on the Palatine Hill, showing him the

INTRODUCTION

places where Rome will be built. He gives to Aeneas to fight on his side at the head of a contingent of Arcadian cavalry his only son, Pallas. They all ride to Agylla to join the Etruscans who have revolted against the tyrant Mezentius. Here Venus gives Aeneas, her son, magical arms and a shield on which are worked, in metal, scenes of the future history of Rome. In Book IX are told the events in the Trojan camp during Aeneas' absence, and in Book X the struggle between Trojans and native Latins continues. Pallas is slain by Turnus in single combat after many brave deeds of his own on the field. Turnus strips from his dead body a belt studded with gold. In Book XI a funeral procession of mourners and warriors escorts the corpse of Pallas back to Pallanteum. Evander laments bitterly for the loss of his only son, grieving that he could not have taken his place. In Book XII, the last of the *Aeneid*, Aeneas and Turnus meet in single combat before the walls of Laurentum. Aeneas has thoughts of sparing Turnus' life when he falls, but at the sight of Pallas' belt which he is wearing as a trophy his anger is aroused, and he slays him. So the war ends with the Trojans victorious.

(iii) *Pallas*

Another title for this book might have been 'The Belt Inlaid with Gold'. It is a tragedy interwoven in the epic narrative which can be seen as a whole only when, as it were, the golden thread is traced running through the great tapestry of the poem. The events which combine to make it one of the greatest stories of ancient literature occur in the last books of the *Aeneid*, in VIII, X, XI and XII. Pallas, the young and tender son of the old king Evander, sets out, bright as a star, at his father's bidding to fight at Aeneas' side. He is all too trusting, an untried tiro, full of youth's enthusiasm, one who will not turn away from a savage foeman older and stronger than himself. After daring exploits on the field when Trojan and Etruscan side by side go out to meet the native Latins, Pallas confronts Turnus. After but a brief, hopeless encounter, Pallas falls his victim. Turnus covetously strips from the pitiful corpse the belt inlaid with gold which Pallas wore with such pride, and triumphantly wears it as his own. From this comes his own undoing. When Aeneas and Turnus at last meet face to face in the single combat which is to end the war, Aeneas, moved to pity, thinks at first to spare his victim's life, but his eyes light on Pallas' belt. Roused to anger when he recalls how Turnus slew Pallas, he deals the death-blow. If Turnus had only shown mercy to the young Pallas, he might himself have obtained it.

This tragic drama, played out through the books of the *Aeneid*, is as great as any masterpiece ever written in ancient or modern literature, and, like the great dramatic tragedies of all ages, touches the depths of human grief. The eternal elements of the hope and promise of youth, the piteous sorrow of an aged father bereft of his only son, are there; there too, the short-lived triumph, the working of revenge, the final penalty paid to the full, which belong to great themes in literature—Evander's agony of lament for Pallas is that of every father for every son lost in war.

INTRODUCTION

The story of Pallas is set against a pre-Roman background, but on the soil where later Rome was to arise. In Vergil's thought the cult of Hercules at the *ara maxima* on the Tiber's bank, practised in his own day and all through the history of Rome, was older than Rome: consequently it could be given a place in his epic. Vergil describes faithfully the essentials of the ritual, and the sacrifice as he must have seen it for himself, although he introduces details from an earlier age, especially in the two priestly families who administer the cult, to give a touch of the antiquity which he loved. We do indeed see one of the great festivals of Rome enacted before our eyes. When Evander leads Aeneas up to his home on the Palatine Hill they pass along valleys, climb grass-grown slopes and hills, and look at sacred places which are to be the heart of future Rome. Vergil shows us the earth and soil of the city which is to be. Evander lives on the hill where the founder of Rome will make the first planting of the infant city. All this is essentially Roman, although Vergil shows us the landscape of an age before Rome was dreamed of.

The theme behind the story of Pallas occurs more than once in the *Aeneid*, but not elsewhere is more finely expressed the heroic sacrifice of young manhood for a noble cause. In Book IX Nisus and Euryalus die in trying to do an act of heroism, in X Lausus the young son of Mezentius the tyrant of Agylla is slain in battle but in the death of Pallas and the mourning of Evander weighed down with helpless old age are to be recognized some of the greatest passages in the whole of the *Aeneid*. Evander's farewell to Pallas (VIII, 560–

83) when he goes out to war, the description of Pallas as he lies dead on the bier (XI, 59–71), and most of all Evander's lament (XI, 152–81) belong to the heights of Vergil's achievement. These passages, which are among the finest not only in his but in any Latin poetry, should be read through again and again until their deep pathos is felt and Vergil's complete mastery of words and of sound is realized.

(iv) *The hexameter*

All Vergil's poems are written in 'hexameters', containing six metrical 'feet' which are either 'dactyls' ($-\cup\cup$) or 'spondees' ($--$), though the last syllable of the line can be short, thus making the last 'spondee' into a 'trochee' ($-\cup$). There is always a break between two words, called a 'caesura', in the middle of the third foot (sometimes only in the fourth foot) which should be marked with a double line or by some other method. The 'caesura' serves to avoid monotony which might otherwise creep in, by interlacing the metrical feet with the words. The scheme therefore is:

$$
\left.\begin{array}{c} -\cup\cup \\ -- \end{array}\right| \left.\begin{array}{c} -\cup\cup \\ -- \end{array}\right| \left.\begin{array}{c} - \\ - \end{array}\right\| \left.\begin{array}{c} \cup\cup \\ - \end{array}\right| \left.\begin{array}{c} -\cup\cup \\ -- \end{array}\right| \begin{array}{c} -\cup\cup \\ \end{array} \left|\begin{array}{c} \cup \\ - \end{array}\right.
$$

You should learn the following six rules (there are more, but these are the easiest and most important):

1. A word ending in a vowel or *-m* which is followed by a word beginning with a vowel (or *h*, which does not count as a consonant) has its last syllable cut off or 'elided', so that it does not count at all

INTRODUCTION

in scansion. But *i* in words like *iam* and *iaceo* is a consonant.

2. All diphthongs, i.e. combinations of two vowels that are pronounced together, like the *ae* in *mensae*, are long (the *u* that always follows *q* does not count as part of a diphthong, e.g. the enclitic -*que* is short).

3. A vowel before two consonants, either in the same word or with one at the end of one word and another at the beginning of the next, forms a long syllable. If *l* or *r* is the second consonant (in the same word) the syllable can sometimes be either short or long, e.g. pătrem. Double *l*, double *r*, *x*, and *z* always make the preceding syllable long, and *h* is ignored.

4. A vowel before another vowel in the same word, when pronounced separately, is usually short, e.g. rĕī.

5. The final -*a* of the first declension ablative singular is long; most other final -*a*'s are short.

6. Final -*i* and -*o* are usually long.

When you start scanning a hexameter first of all look for any elisions according to rule 1 and put brackets round the elided vowel or vowel followed by -*m*. Then mark off the last five syllables, which are always $-\cup\cup\,|\,-\,\cup$, like 'músic of Hómer'. Next, count up the remaining syllables; if there are twelve, there will be four dactyls; if eleven, three dactyls and a spondee; if ten, two of each; if nine, three spondees and a dactyl; if eight, four spondees. Then mark off all the syllables which you know according to the other rules, remembering that a syllable between two longs must itself be long. There will be some that you will not know for certain, but your reading aloud of the line may help. Finally, mark the caesura in the third

INTRODUCTION

(sometimes only in the fourth) foot. Here is an example showing both elision and caesura:

ēxspēct|ātĕ sŏ|lō || Lāu|rēnt(i) ār|vīsquĕ Lă|tīnīs
Aen. VIII, 38

A great Latin scholar, Mackail, has said of Vergil's treatment of the hexameter, 'He stands out as having achieved the utmost beauty, melody, and significance of which human words seem to be capable', and a great poet Tennyson called him

> Wielder of the stateliest measure
> Ever moulded by the lips of man.

To be fully appreciated and understood, Vergil's poetry should be read aloud.

(v) *Reading and translating Vergil*

Before beginning to translate a passage into English, the student should read the Latin through, looking at the structure as a whole, noticing case-endings, agreements, the placing of the verbs, and so on. He could with advantage hear it read aloud by the teacher. In this way he should try to realize something of the meaning before arriving at an English translation.

We cannot fully appreciate Latin poetry unless we learn to think in the Latin order of words which is essentially different from our own. For this reason we should try as far as possible to preserve the Latin order in translation so that the correct emphasis is preserved and the fine-drawn nuances of meaning are not lost. This method of translation has been followed in the notes. When the meaning of a given passage has been

INTRODUCTION

mastered, it is helpful to hear the Latin read aloud. The student should first listen with the Latin text before him and then for a second time, without the text, trying to follow the Latin without translating it, *even in thought*, into English. He should try to master that feeling of 'suspense' which is inherent in Latin in which the verb, the key to the meaning of the sentence, is almost always at the end. It is possible to reach the heart of Vergil's poetry only when we school ourselves to think in the same linguistic processes as those which were natural to a Roman.

THE STORY OF PALLAS

(*Aeneid* VIII, 26–183, 280–369, 454–607)

An outline of the *Aeneid* is given in the Introduction, pp. xvii–xix. The events in Books VII and VIII which lead up to the story of Pallas are as follows:

At the end of their long voyage from Troy, the Trojans reached the mouth of the Tiber, turned their ships upstream, and came to land on the grassy bank. At that very time a rumour was abroad among the native inhabitants of that country that a foreign prince was soon to arrive. Their king, Latinus, who lived in the city of Laurentum about sixteen miles away, had lately agreed to betroth his daughter Lavinia to a native prince, Turnus, chief of the Rutuli. The queen, Amata, her mother, was strongly in favour of the proposed marriage. However, the gods had shown by strange signs that it was against their will, and when Latinus consulted the oracle of Faunus, strange voices in the night warned him that Lavinia was destined, not for Turnus, but for a stranger, and that their descendants would rule the world.

The next day Aeneas sent messengers to Latinus while he himself started to build a camp on the river bank. Latinus welcomed the Trojans, for he recognized in Aeneas the future husband of Lavinia, but the goddess Juno, who hated the Trojans, was determined that there should be war between them and the Latins before the marriage could take place, and opened the Gates of War. Preparations for the coming struggle were made on every side. Native clans assembled before the walls of Laurentum to fight on Turnus' side. Last came Camilla with a troop of Volscian horsemen. Book VIII opens with the signal for war given by Turnus and the rallying of leaders and troops to the call. Aeneas was perplexed, and his thoughts shifted to and fro like light on shimmering water.

THE STORY OF PALLAS [AEN. VIII

26–65. At the dead of night, though still anxious at heart, Aeneas fell asleep on the river bank. As he slept, Tiberinus, the god of the river, dressed in a linen cloak and a wreath of reeds, appeared to him and spoke reassuring words about the future. He told Aeneas that he could be certain of a settled home in the country to which he had come. A white sow with a miraculous litter of thirty young, symbolic of the founding of Alba by Ascanius, would be a sign that his words were true. He urged Aeneas to go and seek the help of King Evander who lived in the city of Pallanteum, and promised to make his journey upstream easy by stilling the current. Then bidding him to make prayer to Juno, he revealed himself as the Tiber, the life-giving stream.

26 Nox erat et terras animalia fessa per omnes
alituum pedudumque genus sopor altus habebat,
cum pater in ripa gelidique sub aetheris axe
Aeneas, tristi turbatus pectora bello,
30 procubuit seramque dedit per membra quietem.
huic deus ipse loci fluvio Tiberinus amoeno
populeas inter senior se attollere frondes
visus (eum tenuis glauco velabat amictu
carbasus, et crines umbrosa tegebat harundo),
35 tum sic adfari et curas his demere dictis:
    'O sate gente deum, Troianam ex hostibus urbem
qui revehis nobis aeternaque Pergama servas,
exspectate solo Laurenti arvisque Latinis,
hic tibi certa domus, certi (ne absiste) penates;
40 neu belli terrere minis; tumor omnis et irae
concessere deum.
iamque tibi, ne vana putes haec fingere somnum,
litoreis ingens inventa sub ilicibus sus
triginta capitum fetus enixa iacebit,
45 alba, solo recubans, albi circum ubera nati.

THE STORY OF PALLAS

[hic locus urbis erit, requies ea certa laborum,]
ex quo ter denis urbem redeuntibus annis
Ascanius clari condet cognominis Albam.
haud incerta cano. nunc qua ratione quod instat
expedias victor, paucis (adverte) docebo.   50
Arcades his oris, genus a Pallante profectum,
qui regem Euandrum comites, qui signa secuti,
delegere locum et posuere in montibus urbem
Pallantis proavi de nomine Pallanteum.
hi bellum adsidue ducunt cum gente Latina;   55
hos castris adhibe socios et foedera iunge.
ipse ego te ripis et recto flumine ducam,
adversum remis superes subvectus ut amnem.
surge age, nate dea, primisque cadentibus astris
Iunoni fer rite preces, iramque minasque   60
supplicibus supera votis. mihi victor honorem
persolves. ego sum pleno quem flumine cernis
stringentem ripas et pinguia culta secantem,
caeruleus Thybris, caelo gratissimus amnis.
hic mihi magna domus, celsis caput urbibus exit.'   65

66–80. After Tiberinus had descended again into the depths of the river, Aeneas, now aroused, made prayers to the spirits of the rivers and streams of the countryside. He promised continued praise and offerings to the Tiber and entreated his help. Then he fitted out two ships for the journey upstream to Pallanteum.

Dixit, deinde lacu fluvius se condidit alto   66
ima petens; nox Aenean somnusque reliquit.
surgit et aetherii spectans orientia solis
lumina rite cavis undam de flumine palmis
sustinet ac tales effundit ad aethera voces: 70

'nymphae, Laurentes nymphae, genus amnibus unde est,
tuque, o Thybri tuo genitor cum flumine sancto,
accipite Aenean et tandem arcete periclis.
quo te cumque lacus miserantem incommoda nostra
75 fonte tenet, quocumque solo pulcherrimus exis,
semper honore meo, semper celebrabere donis,
corniger Hesperidum fluvius regnator aquarum.
adsis o tantum et propius tua numina firmes.'
sic memorat, geminasque legit de classe biremes
80 remigioque aptat, socios simul instruit armis.

81–101. Then came the fulfilment of Tiberinus' prophecy. The Trojans found on the bank among the trees the white sow and her litter, and sacrificed them all to Juno. The flow of the Tiber was miraculously stilled so that they rowed with ease over calm water. After rowing all night along the winding river, at midday on the next day they reached Evander's home.

81   Ecce autem subitum atque oculis mirabile monstrum,
candida per silvam cum fetu concolor albo
procubuit viridique in litore conspicitur sus:
quam pius Aeneas tibi enim, tibi, maxima Iuno,
85 mactat sacra ferens et cum grege sistit ad aram.
Thybris ea fluvium, quam longa est, nocte tumentem
leniit, et tacita refluens ita substitit unda,
mitis ut in morem stagni placidaeque paludis
sterneret aequor aquis, remo ut luctamen abesset.
90 ergo iter inceptum celerant. rumore secundo
labitur uncta vadis abies, mirantur et undae,
miratur nemus insuetum fulgentia longe
scuta virum fluvio pictasque innare carinas.

olli remigio noctemque diemque fatigant
et longos superant flexus, variisque teguntur 95
arboribus, viridesque secant placido aequore silvas.
sol medium caeli conscenderat igneus orbem
cum muros arcemque procul ac rara domorum
tecta vident, quae nunc Romana potentia caelo
aequavit, tum res inopes Euandrus habebat. 100
ocius advertunt proras urbique propinquant.

102–25. On that very day the Arcadians were keeping the Festival of Hercules in a grove on the river bank. The sudden arrival of the Trojans terrified them, but Pallas gave orders that the feast was to go on, and called out to them to ask who they were and where they came from. Overcome with astonishment when he learned that they were Trojans he took them to speak with his father Evander inside the grove.

Forte die sollemnem illo rex Arcas honorem 102
Amphitryoniadae magno divisque ferebat
ante urbem in luco. Pallas huic filius una,
una omnes iuvenum primi pauperque senatus 105
tura dabant, tepidusque cruor fumabat ad aras.
ut celsas videre rates atque inter opacum
adlabi nemus et tacitis incumbere remis,
terrentur visu subito cunctique relictis
consurgunt mensis. audax quos rumpere Pallas 110
sacra vetat raptoque volat telo obvius ipse,
et procul e tumulo: 'iuvenes, quae causa subegit
ignotas temptare vias? quo tenditis?' inquit.
'qui genus? unde domo? pacemne huc fertis an arma?'
tum pater Aeneas puppi sic fatur ab alta 115
paciferaeque manu ramum praetendit olivae:
'Troiugenas ac tela vides inimica Latinis,
quos illi bello profugos egere superbo

Euandrum petimus. ferte haec et dicite lectos
120 Dardaniae venisse duces socia arma rogantes.'
obstipuit tanto percussus nomine Pallas:
'egredere o quicumque es,' ait 'coramque parentem
adloquere ac nostris succede penatibus hospes.'
excepitque manu dextramque amplexus inhaesit.
125 progressi subeunt luco fluviumque relinquunt.

126–51. Aeneas told Evander that he had come with olive branches in his hands, not through fear of meeting a Greek, but because he trusted in their kinship. Since both he and Evander were descended from Atlas, they were of the same stock, and so he approached him in person to ask his help against an enemy with whom Evander was already at war: one who threatened to conquer all the West, but before whom the Trojans were unafraid.

126 Tum regem Aeneas dictis adfatur amicis:
'optime Graiugenum, cui me Fortuna precari
et vitta comptos voluit praetendere ramos,
non equidem extimui Danaum quod ductor et Arcas
130 quodque a stirpe fores geminis coniunctus Atridis;
sed mea me virtus et sancta oracula divum
cognatique patres, tua terris didita fama,
coniunxere tibi et fatis egere volentem.
Dardanus, Iliacae primus pater urbis et auctor,
135 Electra, ut Grai perhibent, Atlantide cretus,
advehitur Teucros; Electram maximus Atlas
edidit, aetherios umero qui sustinet orbes.
vobis Mercurius pater est, quem candida Maia
Cyllenae gelido conceptum vertice fudit;
140 at Maiam, auditis si quicquam credimus, Atlas,
idem Atlas generat caeli qui sidera tollit.
sic genus amborum scindit se sanguine ab uno.

PLATE I. (*a*) A representation of the god of the Tiber, Tiberinus, from a relief on an altar of A.D. 124, found at Ostia, the port of ancient Rome. Dressed in a thin linen cloak, he reclines on an amphora, which symbolises water, and holds a tall reed. See *Aen.* VIII, 31–67, and Background Notes, p. 118.

PLATE I. (*b*) A view of the Tiber, looking down-stream from the bridge situated near where was the *Pons Sublicius*. The Aventine Hill is on the left. The place where Aeneas landed is not in the photograph, but was near the hill on its northern side. See *Aen.* VIII, 97–125, and Background Notes, pp. 122–3.

PLATE II. A view across the Roman Forum of the western corner of the Palatine Hill (*Mons Palatinus*). Somewhere at the foot, perhaps towards the south, was situated the *Lupercal*, the sacred cave connected with the childhood of Romulus and Remus. Archaeological remains of settlers belonging to the early Iron Age have been found on the summit of this edge of the hill. See *Aen*. VIII, 51-4, 343, and Background Notes, pp. 121-2, 130.

PLATE III. A view of the Roman Forum, taken from the Capitoline Hill (*Capitolium*, or *Mons Capitolinus*). See *Aen.* VIII, 361, and Background Notes, p. 132.

PLATE IV. Remains of the paved road (*Clivus Capitolinus*) which in Roman times wound up from the Roman Forum to the summit of the Capitoline Hill to the place where stood the Temple of Jupiter Capitolinus. See *Aen.* VIII, 347-54, and Background Notes, pp. 131-2.

PLATE V. The façade of the Temple of Saturn on the lowest slopes of the Capitoline Hill and on the edge of the Roman Forum. The architecture which still remains belongs to Imperial times, but in origin the Temple probably goes back to the beginning of the Republic. See *Aen.* VIII, 355–8, and Background Notes, pp. 131–2.

PLATE VI. View from the left bank of the Tiber over the open space where was in Roman times the *Forum Boarium*. In the background is the edge of the Palatine Hill. To the right, but not in the photograph, is the place where stood the *ara maxima*, and where the rites of Hercules were observed. See *Aen.* VIII, 102–305, and Background Notes, pp. 123–6.

PLATE VII. View of the southern end of the Capitoline Hill. On this side, but probably more to the east, was the *rupes Tarpeia*, a place of execution, down which criminals were thrown to death. On the summit is the site of the Temple of Jupiter Capitolinus. See *Aen.* VIII, 347-54, and Background Notes, pp. 130-1.

PLATE VIII. Cast of a restored relief, probably of late republican date, from the Basilica Aemilia in the Roman Forum, illustrating the punishment of Tarpeia. The Sabine king, Titus Tatius, looks on while two soldiers batter her to death with their shields. See *Aen.* VIII, 347, and Background Notes, pp. 130–1.

THE STORY OF PALLAS

his fretus non legatos neque prima per artem
temptamenta tui pepigi; me, me ipse meumque
obieci caput et supplex ad limina veni. 145
gens eadem, quae te, crudeli Daunia bello
insequitur; nos si pellant nihil afore credunt
quin omnem Hesperiam penitus sua sub iuga mittant,
et mare quod supra teneant quodque adluit infra.
accipe daque fidem. sunt nobis fortia bello 150
pectora, sunt animi et rebus spectata iuventus.'

152–74. Recognizing Aeneas' likeness to his father Anchises, Evander received him with joy. He remembered in his early manhood King Priam passing through Arcadia on his way to Salamis, and how among his companions he had seen and admired Anchises. They became friends and on his departure Anchises had given him generous gifts which now belonged to Pallas. He promised Aeneas his support and invited him in the mean time to join the feast.

Dixerat Aeneas. ille os oculosque loquentis 152
iamdudum et totum lustrabat lumine corpus.
tum sic pauca refert: 'ut te, fortissime Teucrum,
accipio agnoscoque libens! ut verba parentis 155
et vocem Anchisae magni vultumque recordor!
nam memini Hesionae visentem regna sororis
Laomedontiaden Priamum Salamina petentem
protinus Arcadiae gelidos invisere fines.
tum mihi prima genas vestibat flore iuventas, 160
mirabarque duces Teucros, mirabar et ipsum
Laomedontiaden; sed cunctis altior ibat
Anchises. mihi mens iuvenali ardebat amore
compellare virum et dextrae coniungere dextram;
accessi et cupidus Phenei sub moenia duxi. 165
ille mihi insignem pharetram Lyciasque sagittas

discedens chlamydemque auro dedit intertextam,
frenaque bina meus quae nunc habet aurea Pallas.
ergo et quam petitis iuncta est mihi foedere dextra,
170 et lux cum primum terris se crastina reddet,
auxilio laetos dimittam opibusque iuvabo.
interea sacra haec, quando huc venistis amici,
annua, quae differre nefas, celebrate faventes
nobiscum, et iam nunc sociorum adsuescite mensis.'

175–83. Evander ordered the feast to be resumed. Aeneas was given a seat of honour on a chair covered with a lion-skin. The attendants served out the feast and set before Aeneas and the Trojans a generous share of the victims' flesh.

175   Haec ubi dicta, dapes iubet et sublata reponi
pocula gramineoque viros locat ipse sedili,
praecipuumque toro et villosi pelle leonis
accipit Aenean solioque invitat acerno.
tum lecti iuvenes certatim araeque sacerdos
180 viscera tosta ferunt taurorum, onerantque canistris
dona laboratae Cereris, Bacchumque ministrant.
vescitur Aeneas simul et Troiana iuventus
perpetui tergo bovis et lustralibus extis.

184–279 (omitted). When the feasting was ended, Evander told the tale of Hercules and Cacus to the assembled company to explain to them that the rites had been established there, not because of idle superstition, but to do honour to Hercules as their liberator. Pointing to some shattered rocks on the edge of the Aventine Hill, he told how once a cave was there in which lived a man-eating, fire-breathing giant who was a danger to all the countryside until Hercules came that way. He was returning from Spain with the cattle of the three-bodied monster Geryon whom he had slain: Cacus

stole four bulls and four heifers, pulling them backwards by
their tails into the cave, so that their tracks did not betray
his theft. In the morning, however, when Hercules was
driving the cattle on, they lowed, and one answered from
inside the cave. In great anger Hercules seized his club and
leapt up the hillside. Cacus took refuge inside and pulled a
great rock across the entrance, bolting and barring himself
in. When he could find no other means of entry, Hercules
pulled down from its foundations a great rock, which hung
over the cave's mouth. Its fall shook the river bank and laid
open Cacus' hiding-place. With no hope of escape Cacus
tried the ruse of belching out fire and smoke, but Hercules
leapt through the rolling fumes and caught and strangled
him so that he lay stretched dead at his feet. For this
exploit the annual sacrifice was celebrated and the rites
observed.

280–305. Evander's tale ended, in the evening the feast
was renewed. The priests brought flares to light the scene.
Then the Salii danced around the altar, singing in chorus
hymns in praise of Hercules. They told of his strength when
only a child, how in manhood he sacked cities, and then of
his labours, the slaying of the centaurs and the Nemean lion;
how he carried off from the underworld the dog Cerberus,
and slew Typhoeus and the Lernaean hydra. Especially did
they honour him in song for the slaying of Cacus.

Devexo interea propior fit Vesper Olympo.   280
iamque sacerdotes primusque Potitius ibant
pellibus in morem cincti, flammasque ferebant.
instaurant epulas et mensae grata secundae
dona ferunt cumulantque oneratis lancibus aras.
tum Salii ad cantus incensa altaria circum   285
populeis adsunt evincti tempora ramis,
hic iuvenum chorus, ille senum, qui carmine laudes
Herculeas et facta ferunt: ut prima novercae

monstra manu geminosque premens eliserit angues,
290 ut bello egregias idem disiecerit urbes,
Troiamque Oechaliamque, ut duros mille labores
rege sub Eurystheo fatis Iunonis iniquae
pertulerit. 'tu nubigenas, invicte, bimembres,
Hylaeumque Pholumque, manu, tu Cresia mactas
295 prodigia et vastum Nemeae sub rupe leonem.
te Stygii tremuere lacus, te ianitor Orci
ossa super recubans antro semesa cruento;
nec te ullae facies, non terruit ipse Typhoeus
arduus arma tenens; non te rationis egentem
300 Lernaeus turba capitum circumstetit anguis.
salve, vera Iovis proles, decus addite divis,
et nos et tua dexter adi pede sacra secundo.'
talia carminibus celebrant; super omnia Caci
speluncam adiciunt spirantemque ignibus ipsum.
305 consonat omne nemus strepitu collesque resultant.

306–36. When the feasting was ended, Evander led Aeneas and Iulus to Pallanteum. Aeneas was fascinated by all that he saw and especially by the memorials of earlier men. As they went along, Evander told them of earlier ages, of the first men born of trees, who had no knowledge of agriculture: then of the golden age ushered in by Saturn, who gave the name Latium to the land; lastly of a degenerate age when greed crept in and kings ruled and when the Tiber lost its ancient name. It was then that he himself had founded Pallanteum at the bidding of his mother.

306 Exim se cuncti divinis rebus ad urbem
perfectis referunt. ibat rex obsitus aevo,
et comitem Aenean iuxta natumque tenebat
ingrediens varioque viam sermone levabat.
310 miratur facilesque oculos fert omnia circum

AEN. VIII] THE STORY OF PALLAS

Aeneas, capiturque locis et singula laetus
exquiritque auditque virum monimenta priorum.
tum rex Euandrus Romanae conditor arcis:
'haec nemora indigenae Fauni Nymphaeque tenebant
gensque virum truncis et duro robore nata, 315
quis neque mos neque cultus erat, nec iungere tauros
aut componere opes norant aut parcere parto,
sed rami atque asper victu venatus alebat.
primus ab aetherio venit Saturnus Olympo
arma Iovis fugiens et regnis exsul ademptis. 320
is genus indocile ac dispersum montibus altis
composuit legesque dedit, Latiumque vocari
maluit, his quoniam latuisset tutus in oris.
aurea quae perhibent illo sub rege fuerunt
saecula: sic placida populos in pace regebat, 325
deterior donec paulatim ac decolor aetas
et belli rabies et amor successit habendi.
tum manus Ausonia et gentes venere Sicanae,
saepius et nomen posuit Saturnia tellus;
tum reges asperque immani corpore Thybris, 330
a quo post Itali fluvium cognomine Thybrim
diximus; amisit verum vetus Albula nomen.
me pulsum patria pelagique extrema sequentem
Fortuna omnipotens et ineluctabile fatum
his posuere locis, matrisque egere tremenda 335
Carmentis Nymphae monita et deus auctor Apollo.'

337–69. As he spoke of Carmentis, his mother, Evander showed Aeneas her altar and the gate named after her and told of her prophecies. Then as they walked on he pointed out places which were to be famous in the future, the Asylum, the Lupercal, the Argiletum, the Capitoline Hill covered with forest growth but already the haunt of Jupiter, and the ruins of villages belonging to an earlier age. At last they came to Evander's house, and saw cattle pastured in the Roman forum and the Carinae. Evander welcomed the Trojans to his home, saying that although it was but a humble dwelling it had once given hospitality to a god.

337 Vix ea dicta, dehinc progressus monstrat et aram
et Carmentalem Romani nomine portam
quam memorant, Nymphae priscum Carmentis honorem,
340 vatis fatidicae, cecinit quae prima futuros
Aeneadas magnos et nobile Pallanteum.
hinc lucum ingentem, quem Romulus acer asylum
rettulit, et gelida monstrat sub rupe Lupercal
Parrhasio dictum Panos de more Lycaei.
345 nec non et sacri monstrat nemus Argileti
testaturque locum et letum docet hospitis Argi.
hinc ad Tarpeiam sedem et Capitolia ducit
aurea nunc, olim silvestribus horrida dumis.
iam tum religio pavidos terrebat agrestes
350 dira loci, iam tum silvam saxumque tremebant.
'hoc nemus, hunc' inquit 'frondoso vertice collem
(quis deus incertum est) habitat deus; Arcades ipsum
credunt se vidisse Iovem, cum saepe nigrantem
aegida concuteret dextra nimbosque cieret.
355 haec duo praeterea disiectis oppida muris,
reliquias veterumque vides monimenta virorum.
hanc Ianus pater, hanc Saturnus condidit arcem;
Ianiculum huic, illi fuerat Saturnia nomen.'
talibus inter se dictis ad tecta subibant
360 pauperis Euandri, passimque armenta videbant
Romanoque foro et lautis mugire Carinis.
ut ventum ad sedes, 'haec' inquit 'limina victor

THE STORY OF PALLAS

Alcides subiit, haec illum regia cepit.
aude, hospes, contemnere opes et te quoque dignum
finge deo, rebusque veni non asper egenis.' 365
dixit, et angusti subter fastigia tecti
ingentem Aenean duxit stratisque locavit
effultum foliis et pelle Libystidis ursae:
nox ruit et fuscis tellurem amplectitur alis.

370–453 (omitted). Venus, alarmed for her son Aeneas, begged Vulcan the god of fire to forge divine armour to aid him in the war. Touched by her charms and her pleading, Vulcan agreed to do all that she asked. Rising very early at the time when a thrifty housewife is busy at the loom, Vulcan went in haste to his forge under Aetna where the Cyclopes were forging a thunderbolt, a chariot for Mars, and an aegis for Pallas. Calling off these tasks, he told them to forge armour for Aeneas with all speed. They turned to this toil with alacrity and began to shape a great shield to be proof against all the weapons of the Latins.

454–68. The next morning rising early, Evander put on his tunic and Etruscan sandals, girt his sword at his side and slung a panther-skin over his shoulders. Calling his two dogs to accompany him he went to converse with Aeneas about means of helping him.

Haec pater Aeoliis properat dum Lemnius oris, 454
Euandrum ex humili tecto lux suscitat alma 455
et matutini volucrum sub culmine cantus.
consurgit senior tunicaque inducitur artus
et Tyrrhena pedum circumdat vincula plantis.
tum lateri atque umeris Tegeaeum subligat ensem
demissa ab laeva pantherae terga retorquens. 460
nec non et gemini custodes limine ab alto
praecedunt gressumque canes comitantur erilem

hospitis Aeneae sedem et secreta petebat
sermonum memor et promissi muneris heros.
465 nec minus Aeneas se matutinus agebat.
filius huic Pallas, illi comes ibat Achates.
congressi iungunt dextras mediisque residunt
aedibus et licito tandem sermone fruuntur.

469–519. Evander told Aeneas that his own resources were too small to help the Trojan cause, and that his own situation in the war was desperate, but that the Etruscans of neighbouring Agylla were looking for a leader. They had lately risen against Mezentius their cruel tyrant, whose atrocities were too dreadful to speak of; he had even tied men to dead bodies and left them to die a lingering and horrible death. At last the citizens surrounded and burnt his house, but he himself escaped and found refuge with Turnus. All Etruria had risen against him in righteous anger and were demanding a leader to bring him to justice. Evander besought Aeneas to be at their head. Ships were indeed ready on the shore; but the seers had warned them that they must seek a foreign leader. An embassy from Tarcho had lately appealed to Evander but he was too infirm with age: and Pallas was half Italian. Aeneas was clearly marked out by heaven for the task. At this Evander offered to send Pallas, his only son, with Aeneas to the war, at the head of a contingent of Arcadian cavalry.

469 Rex prior haec:
470 'maxime Teucrorum ductor, quo sospite numquam
res equidem Troiae victas aut regna fatebor,
nobis ad belli auxilium pro nomine tanto
exiguae vires; hinc Tusco claudimur amni,
hinc Rutulus premit et murum circumsonat armis.
475 sed tibi ego ingentes populos opulentaque regnis

iungere castra paro, quam fors inopina salutem
ostentat. fatis huc te poscentibus adfers.
haud procul hinc saxo incolitur fundata vetusto
urbis Agyllinae sedes, ubi Lydia quondam
gens, bello praeclara, iugis insedit Etruscis.   480
hanc multos florentem annos rex deinde superbo
imperio et saevis tenuit Mezentius armis.
quid memorem infandas caedes, quid facta tyranni
effera? di capiti ipsius generique reservent!
mortua quin etiam iungebat corpora vivis   485
componens manibusque manus atque oribus ora,
tormenti genus, et sanie taboque fluentes
complexu in misero longa sic morte necabat.
at fessi tandem cives infanda furentem
armati circumsistunt ipsumque domumque,   490
obtruncant socios, ignem ad fastigia iactant.
ille inter caedem Rutulorum elapsus in agros
confugere et Turni defendier hospitis armis.
ergo omnis furiis surrexit Etruria iustis,
regem ad supplicium praesenti Marte reposcunt.   495
his ego te, Aenea, ductorem milibus addam.
toto namque fremunt condensae litore puppes
signaque ferre iubent, retinet longaevus haruspex
fata canens: "o Maeoniae delecta iuventus,
flos veterum virtusque virum, quos iustus in hostem   500
fert dolor et merita accendit Mezentius ira,
nulli fas Italo tantam subiungere gentem:
externos optate duces." tum Etrusca resedit
hoc acies campo monitis exterrita divum.
ipse oratores ad me regnique coronam   505
cum sceptro misit mandatque insignia Tarcho,
succedam castris Tyrrhenaque regna capessam.

sed mihi tarda gelu saeclisque effeta senectus
invidet imperium seraeque ad fortia vires.
510 natum exhortarer, ni mixtus matre Sabella
hinc partem patriae traheret. tu, cuius et annis
et generi fata indulgent, quem numina poscunt,
ingredere, o Teucrum atque Italum fortissime ductor.
hunc tibi praeterea, spes et solacia nostri,
515 Pallanta adiungam; sub te tolerare magistro
militiam et grave Martis opus, tua cernere facta
adsuescat, primis et te miretur ab annis.
Arcadas huic equites bis centum, robora pubis
lecta dabo, totidemque suo tibi munere Pallas.'

520–40. At Evander's words sad thoughts would have taken possession of their minds if Venus had not sent them a sign. Suddenly lightning flashed and there was the sound of a trumpet. They saw a vision of arms in the clear air. Aeneas recognized it as that of the arms made by Vulcan once promised him by his mother Venus. He knew what havoc the coming war was to bring, and how many were to drown in the Tiber's stream.

520 Vix ea fatus erat, defixique ora tenebant
Aeneas Anchisiades et fidus Achates,
multaque dura suo tristi cum corde putabant,
ni signum caelo Cytherea dedisset aperto.
namque improviso vibratus ab aethere fulgor
525 cum sonitu venit et ruere omnia visa repente,
Tyrrhenusque tubae mugire per aethera clangor.
suspiciunt, iterum atque iterum fragor increpat ingens.
arma inter nubem caeli in regione serena
per sudum rutilare vident et pulsa tonare.
530 obstipuere animis alii, sed Troius heros
agnovit sonitum et divae promissa parentis.

AEN. VIII]   THE STORY OF PALLAS

tum memorat: 'ne vero, hospes, ne quaere profecto
quem casum portenta ferant: ego poscor. Olympo
hoc signum cecinit missuram diva creatrix,
si bellum ingrueret, Volcaniaque arma per auras   535
laturam auxilio.
heu quantae miseris caedes Laurentibus instant!
quas poenas mihi, Turne, dabis! quam multa sub
    undas
scuta virum galeasque et fortia corpora volves,
Thybri pater! poscant acies et foedera rumpant.'   540

541–53. Aeneas offered sacrifice to Hercules and the Penates. Then he visited the boats and chose from the number of his company those who were to follow him in war: the rest he sent back with news to Iulus. The Trojans who were to ride to Agylla were given horses; that of Aeneas was covered with a lion-skin with gilded claws.

Haec ubi dicta dedit, solio se tollit ab alto   541
et primum Herculeis sopitas ignibus aras
excitat, hesternumque larem parvosque penates
laetus adit; mactat lectas de more bidentes
Euandrus pariter, pariter Troiana iuventus.   545
post hinc ad naves graditur sociosque revisit,
quorum de numero qui sese in bella sequantur
praestantes virtute legit; pars cetera prona
fertur aqua segnisque secundo defluit amni,
nuntia ventura Ascanio rerumque patrisque.   550
dantur equi Teucris Tyrrhena petentibus arva;
ducunt exsortem Aeneae, quem fulva leonis
pellis obit totum praefulgens unguibus aureis.

554–84. A rumour that spread panic went about that horsemen were speeding to the gates of Agylla. Evander took leave of Pallas, longing for the years that were past when he had piled up the shields of those whom he slew at Praeneste and slain the monster Erulus. He would not have let his own son go alone nor have received insults from Mezentius if he had had his old strength. He prayed that his life might be prolonged if Pallas were to return safe: if this could not be, he longed for instant death. He was carried fainting into his house.

554 Fama volat parvam subito vulgata per urbem
555 ocius ire equites Tyrrheni ad limina regis.
vota metu duplicant matres, propiusque periclo
it timor et maior Martis iam apparet imago.
tum pater Euandrus dextram complexus euntis
haeret inexpletus lacrimans ac talia fatur:
560 'o mihi praeteritos referat si Iuppiter annos,
qualis eram cum primam aciem Praeneste sub ipsa
stravi scutorumque incendi victor acervos
et regem hac Erulum dextra sub Tartara misi,
nascenti cui tres animas Feronia mater
565 (horrendum dictu) dederat, terna arma movenda
(ter leto sternendus erat; cui tum tamen omnes
abstulit haec animas dextra et totidem exuit armis):
non ego nunc dulci amplexu divellerer usquam,
nate, tuo, neque finitimo Mezentius umquam
570 huic capiti insultans tot ferro saeva dedisset
funera, tam multis viduasset civibus urbem.
at vos, o superi, et divum tu maxime rector
Iuppiter, Arcadii, quaeso, miserescite regis
et patrias audite preces: si numina vestra
575 incolumem Pallanta mihi, si fata reservant,
si visurus eum vivo et venturus in unum:
vitam oro, patior quemvis durare laborem.

sin aliquem infandum casum, Fortuna, minaris,
nunc, nunc o liceat crudelem abrumpere vitam,
dum curae ambiguae, dum spes incerta futuri, 580
dum te, care puer, mea sola et sera voluptas,
complexu teneo, gravior neu nuntius aures
vulneret.' haec genitor digressu dicta supremo
fundebat: famuli conlapsum in tecta ferebant.

585–607. The Trojans and Arcadians, with Pallas beautiful as the morning star, rode out to Agylla. Mothers watched in terror from the walls. Following a track through the brushwood, they came to a grove near Caere's stream, which, it was said, men of old had consecrated to Silvanus the country god. Near here Tarcho and the Etruscans were encamped. Aeneas and the Trojans joined him and rested after their ride.

Iamque adeo exierat portis equitatus apertis 585
Aeneas inter primos et fidus Achates,
inde alii Troiae proceres, ipse agmine Pallas
in medio chlamyde et pictis conspectus in armis,
qualis ubi Oceani perfusus Lucifer unda,
quem Venus ante alios astrorum diligit ignes, 590
extulit os sacrum caelo tenebrasque resolvit.
stant pavidae in muris matres oculisque sequuntur
pulveream nubem et fulgentes aere catervas.
olli per dumos, qua proxima meta viarum,
armati tendunt; it clamor, et agmine facto 595
quadripedante putrem sonitu quatit ungula campum.
est ingens gelidum lucus prope Caeritis amnem,
religione patrum late sacer; undique colles
inclusere cavi et nigra nemus abiete cingunt.
Silvano fama est veteres sacrasse Pelasgos, 600
arvorum pecorisque deo, lucumque diemque,

qui primi fines aliquando habuere Latinos.
haud procul hinc Tarcho et Tyrrheni tuta tenebant
castra locis, celsoque omnis de colle videri
605 iam poterat legio et latis tendebat in arvis.
huc pater Aeneas et bello lecta iuventus
succedunt, fessique et equos et corpora curant.

608–731 (end of *Aeneid* VIII, omitted). Venus brought to Aeneas as he sat in a glade apart the divine armour and arms which Vulcan had made at her bidding. Among them was the great shield decorated with scenes of future Roman history, from the foundation legend of Romulus and Remus which told of the first beginnings of the City, to Augustus' victory at Actium over Antony when he returned in triumph to Rome and received the homage of conquered nations.

The scene changes in Book IX to the Trojan camp at the mouth of the Tiber where Ascanius had been left in charge during Aeneas' absence. Turnus with his Rutulian host seized the opportunity and attacked. Two friends still only boys, Nisus and Euryalus, offered to try and find their way to Pallanteum to take the news to Aeneas, but they were caught and slain in the woods and their heads paraded before the camp. Then a determined assault was made on the walls, during which Ascanius slew Numanus with his bow and arrow: this was his first exploit in war: Turnus forced a way into the camp, and, after dealing havoc among the Trojans, leapt into the Tiber and joined his comrades. In Book X the assault was renewed. At length Aeneas arrived by sea with the Etruscan fleet. The sea-nymph Cymodocea told him of the desperate situation in the camp and sped his ship onwards. When the flash of his shield was seen as a signal, the Trojans were comforted. Turnus attacked as they landed, and a general combat followed.

THE STORY OF PALLAS

(*Aeneid* x, 362–79, 439–509)

362–79. In another part of the fight, the Arcadian cavalry, who had ridden along the Tiber's bank from Pallanteum, had to dismount to pass down a ravine and could not hold their ground on foot. Pallas rallied them in the name of Evander and his own trust in them. He told them that flight could not help them, for the sea was in front, but that they must strike where the enemy was thickest.

At parte ex alia, qua saxa rotantia late 362
impulerat torrens arbustaque diruta ripis,
Arcadas insuetos acies inferre pedestres
ut vidit Pallas Latio dare terga sequaci 365
(aspera quis natura loci dimittere quando
suasit equos), unum quod rebus restat egenis,
nunc prece, nunc dictis virtutem accendit amaris:
'quo fugitis, socii? per vos et fortia facta,
per ducis Euandri nomen devictaque bella 370
spemque meam, patriae quae nunc subit aemula laudi,
fidite ne pedibus. ferro rumpenda per hostes
est via. qua globus ille virum densissimus urget,
hac vos et Pallanta ducem patria alta reposcit.
numina nulla premunt, mortali urgemur ab hoste 375
mortales; totidem nobis animaeque manusque.
ecce maris magna claudit nos obice pontus,
deest iam terra fugae: pelagus Troiamne petemus?'
haec ait, et medius densos prorumpit in hostes.

380–438 (omitted). Pallas fought his victorious way through the enemy's ranks, slaying in combat one hero after another until his own men were fired again for the fight. He slew Halaesus and would have made for Lausus too, but Jupiter did not permit these two to meet in combat, for each was destined to fall beneath a mightier hand.

THE STORY OF PALLAS [AEN. X

439–78. Turnus, at the bidding of his sister, claimed the right to meet Pallas in single combat, expressing the wish that Evander were there to see. Pallas gazed at his great frame unafraid, and came forth into the cleared space. Turnus leapt from his chariot and advanced to the battle like a lion on its prey. Pallas struck first, praying to Hercules that he might strip Turnus of his arms. Jupiter forbade his wish to be granted, reminding Hercules that his own son had perished at Troy. Pallas' spear only grazed Turnus' body through the top of his shield.

439 Interea soror alma monet succedere Lauso
440 Turnum, qui volucri curru medium secat agmen.
ut vidit socios: 'tempus desistere pugnae;
solus ego in Pallanta feror, soli mihi Pallas
debetur; cuperem ipse parens spectator adesset.'
haec ait, et socii cesserunt aequore iusso.
445 at Rutulum abscessu iuvenis tum iussa superba
miratus stupet in Turno corpusque per ingens
lumina volvit obitque truci procul omnia visu,
talibus et dictis it contra dicta tyranni:
'aut spoliis ego iam raptis laudabor opimis
450 aut leto insigni: sorti pater aequus utrique est.
tolle minas.' fatus medium procedit in aequor;
frigidus Arcadibus coit in praecordia sanguis.
desiluit Turnus biiugis, pedes apparat ire
comminus; utque leo, specula cum vidit ab alta
455 stare procul campis meditantem in proelia taurum,
advolat: haud alia est Turni venientis imago.
hunc ubi contiguum missae fore credidit hastae,
ire prior Pallas, si qua fors adiuvet ausum
viribus imparibus, magnumque ita ad aethera fatur:
460 'per patris hospitium et mensas, quas advena adisti,
te precor, Alcide, coeptis ingentibus adsis.

THE STORY OF PALLAS

cernat semineci sibi me rapere arma cruenta
victoremque ferant morientia lumina Turni.'
audiit Alcides iuvenem magnumque sub imo
corde premit gemitum lacrimasque effundit inanes.   465
tum genitor natum dictis adfatur amicis:
'stat sua cuique dies, breve et inreparabile tempus
omnibus est vitae; sed famam extendere factis,
hoc virtutis opus. Troiae sub moenibus altis
tot nati cecidere deum, quin occidit una   470
Sarpedon, mea progenies: etiam sua Turnum
fata vocant metasque dati pervenit ad aevi.'
sic ait, atque oculos Rutulorum reicit arvis.
at Pallas magnis emittit viribus hastam
vaginaque cava fulgentem deripit ensem.   475
illa volans umeri surgunt qua tegmina summa
incidit, atque viam clipei molita per oras
tandem etiam magno strinxit de corpore Turni.

479–509. Then Turnus, poising his spear, struck at Pallas: it pierced the centre of his shield and lodged in his breast. He rolled dying on his wound. Turnus exulted over him but gave back his body for burial; yet he first stripped off a belt inlaid with gold and put it on himself. Little did he think that this trophy would prove to be his own undoing!

Hic Turnus ferro praefixum robur acuto   479
in Pallanta diu librans iacit atque ita fatur: 480
'aspice num mage sit nostrum penetrabile telum.'
dixerat; at clipeum, tot ferri terga, tot aeris,
quem pellis totiens obeat circumdata tauri,
vibranti cuspis medium transverberat ictu
loricaeque moras et pectus perforat ingens.   485
ille rapit calidum frustra de vulnere telum:
una eademque via sanguis animusque sequuntur.

corruit in vulnus (sonitum super arma dedere)
et terram hostilem moriens petit ore cruento.
490 quem Turnus super adsistens:
'Arcades, haec' inquit 'memores mea dicta referte
Euandro: qualem meruit, Pallanta remitto.
quisquis honos tumuli, quidquid solamen humandi est,
largior. haud illi stabunt Aeneia parvo
495 hospitia.' et laevo pressit pede talia fatus
exanimem rapiens immania pondera baltei
impressumque nefas: una sub nocte iugali
caesa manus iuvenum foede thalamique cruenti,
quae Clonus Eurytides multo caelaverat auro;
500 quo nunc Turnus ovat spolio gaudetque potitus.
nescia mens hominum fati sortisque futurae
et servare modum rebus sublata secundis!
Turno tempus erit magno cum optaverit emptum
intactum Pallanta, et cum spolia ista diemque
505 oderit. at socii multo gemitu lacrimisque
impositum scuto referunt Pallanta frequentes.
o dolor atque decus magnum rediture parenti,
haec te prima dies bello dedit, haec eadem aufert,
cum tamen ingentes Rutulorum linquis acervos!

The rest of Book x, 510–908, is omitted. It relates how Aeneas forced his way through the enemy and gained the Trojan camp where Ascanius, who had held it during his absence, received him joyfully. The fight became general again; in its course Aeneas killed Lausus, Mezentius' young son, but in pity refrained from spoiling the body of its arms. Mezentius the tyrant of Agylla challenged Aeneas but was slain by him.

THE STORY OF PALLAS

(Aeneid xi, 24–99, 139–81)*

Book XI opens with the trophy on which Aeneas displayed the arms he took from Mezentius. He told his men that since the tyrant was now dead, their next task was to attack the walls of Laurentum, but first they had to bury their dead.

24–58. Aeneas bade his men prepare to escort home the body of Pallas, then he returned to his tent where Acoetes stood beside the bier and all around were the mourners: the sounds of lamentation increased. Aeneas, moved to weeping, grieved that Fortune had grudged him Pallas. It was not such a promise as this that he had given Evander when he let him go to face a people hard in battle. Perhaps he was even now praying for his safe return. Yet Evander would see wounds of which he could be proud. What a heavy loss to the Trojan cause was the death of Pallas!

'Ite' ait, 'egregias animas, quae sanguine nobis   24
hanc patriam peperere suo, decorate supremis   25
muneribus, maestamque Euandri primus ad urbem
mittatur Pallas, quem non virtutis egentem
abstulit atra dies et funere mersit acerbo.'
    Sic ait inlacrimans, recipitque ad limina gressum
corpus ubi exanimi positum Pallantis Acoetes   30
servabat senior, qui Parrhasio Euandro
armiger ante fuit, sed non felicibus aeque
tum comes auspiciis caro datus ibat alumno.
circum omnis famulumque manus Troianaque turba
et maestum Iliades crinem de more solutae.   35
ut vero Aeneas foribus sese intulit altis
ingentem gemitum tunsis ad sidera tollunt
pectoribus, maestoque immugit regia luctu.
ipse caput nivei fultum Pallantis et ora

40 ut vidit levique patens in pectore vulnus
cuspidis Ausoniae, lacrimis ita fatur obortis:
'tene' inquit, 'miserande puer, cum laeta veniret,
invidit Fortuna mihi, ne regna videres
nostra neque ad sedes victor veherere paternas?
45 non haec Euandro de te promissa parenti
discedens dederam, cum me complexus euntem
mitteret in magnum imperium metuensque moneret
acres esse viros, cum dura proelia gente.
et nunc ille quidem spe multum captus inani
50 fors et vota facit cumulatque altaria donis,
nos iuvenem exanimum et nil iam caelestibus ullis
debentem vano maesti comitamur honore.
infelix, nati funus crudele videbis!
hi nostri reditus exspectatique triumphi?
55 haec mea magna fides? at non, Euandre, pudendis
vulneribus pulsum aspicies, nec sospite dirum
optabis nato funus pater. hei mihi, quantum
praesidium, Ausonia, et quantum tu perdis, Iule!'

59–99. A thousand men were chosen to escort the dead Pallas laid on a rustic bier and covered with rich robes. He was as fair as a newly-plucked flower left to die. A long procession of spoils and captives followed and then came Acoetes and Pallas' chariot and war-horse dropping tears. Aeneas uttered the last 'Farewell' to the dead as the procession passed.

59   Haec ubi deflevit, tolli miserabile corpus
60 imperat, et toto lectos ex agmine mittit
mille viros qui supremum comitentur honorem
intersintque patris lacrimis, solacia luctus
exigua ingentis, misero sed debita patri.
haud segnes alii crates et molle feretrum

THE STORY OF PALLAS

arbuteis texunt virgis et vimine querno 65
exstructosque toros obtentu frondis inumbrant.
hic iuvenem agresti sublimem stramine ponunt:
qualem virgineo demessum pollice florem
seu mollis violae seu languentis hyacinthi,
cui neque fulgor adhuc nec dum sua forma recessit, 70
non iam mater alit tellus viresque ministrat.
tum geminas vestes auroque ostroque rigentes
extulit Aeneas, quas illi laeta laborum
ipsa suis quondam manibus Sidonia Dido
fecerat et tenui telas discreverat auro. 75
harum unam iuveni supremum maestus honorem
induit arsurasque comas obnubit amictu,
multaque praeterea Laurentis praemia pugnae
aggerat et longo praedam iubet ordine duci;
addit equos et tela quibus spoliaverat hostem. 80
vinxerat et post terga manus, quos mitteret umbris
inferias, caeso sparsurus sanguine flammas,
indutosque iubet truncos hostilibus armis
ipsos ferre duces inimicaque nomina figi.
ducitur infelix aevo confectus Acoetes, 85
pectora nunc foedans pugnis, nunc unguibus ora,
sternitur et toto proiectus corpore terrae.
ducunt et Rutulo perfusos sanguine currus.
post bellator equus positis insignibus Aethon
it lacrimans guttisque umectat grandibus ora. 90
hastam alii galeamque ferunt, nam cetera Turnus
victor habet. tum maesta phalanx Teucrique
    sequuntur
Tyrrhenique omnes et versis Arcades armis.
postquam omnis longe comitum praecesserat ordo,
substitit Aeneas gemituque haec addidit alto: 95

'nos alias hinc ad lacrimas eadem horrida belli
fata vocant: salve aeternum mihi, maxime Palla,
aeternumque vale.' nec plura effatus ad altos
tendebat muros gressumque in castra ferebat.

100–38 (omitted). A truce was arranged between the Trojans and Latins for six days.

139–81. The news reached Evander and a troop of mourners went out to meet the procession. Evander lamented most bitterly over Pallas' body, for he had known all along what the end would be. He the aged father had now outlived his son; but his sorrow was tempered with the pride he could not help but feel in the trophies of war which were signs of Pallas' brave fighting. His only remaining desire was that his son's death might be avenged by Aeneas.

139 Et iam Fama volans, tanti praenuntia luctus,
140 Euandrum Euandrique domos et moenia replet,
quae modo victorem Latio Pallanta ferebat.
Arcades ad portas ruere et de more vetusto
funereas rapuere faces; lucet via longo
ordine flammarum et late discriminat agros.
145 contra turba Phrygum veniens plangentia iungit
agmina. quae postquam matres succedere tectis
viderunt, maestam incendunt clamoribus urbem
at non Euandrum potis est vis ulla tenere,
sed venit in medios. feretro Pallante reposto
150 procubuit super atque haeret lacrimansque
    gemensque,
et via vix tandem vocis laxata dolore est:
'non haec, o Palla, dederas promissa parenti.
cautius ut saevo velles te credere Marti!
haud ignarus eram quantum nova gloria in armis

et praedulce decus primo certamine posset. 155
primitiae iuvenis miserae bellique propinqui
dura rudimenta, et nulli exaudita deorum
vota precesque meae! tuque, o sanctissima coniunx,
felix morte tua neque in hunc servata dolorem!
contra ego vivendo vici mea fata, superstes 160
restarem ut genitor. Troum socia arma secutum
obruerent Rutuli telis! animam ipse dedissem
atque haec pompa domum me, non Pallanta, referret!
nec vos arguerim, Teucri, nec foedera nec quas
iunximus hospitio dextras: sors ista senectae 165
debita erat nostrae. quod si immatura manebat
mors natum, caesis Volscorum milibus ante
ducentem in Latium Teucros cecidisse iuvabit.
quin ego non alio digner te funere, Palla,
quam pius Aeneas et quam magni Phryges et quam 170
Tyrrhenique duces, Tyrrhenum exercitus omnis.
magna tropaea ferunt quos dat tua dextera leto;
tu quoque nunc stares immanis truncus in armis,
esset par aetas et idem si robur ab annis,
Turne. sed infelix Teucros quid demoror armis? 175
vadite et haec memores regi mandata referte:
quod vitam moror invisam Pallante perempto
dextera causa tua est, Turnum natoque patrique
quam debere vides. meritis vacat hic tibi solus
fortunaeque locus. non vitae gaudia quaero, 180
nec fas, sed nato manes perferre sub imos.'

The remainder of Book XI, 182-915, is omitted. Latinus attempted to make peace, realizing that the Latins were fighting against a race favoured by heaven. While a bitter debate went on in the council-chamber between Drances and Turnus, Aeneas moved his troops from the Tiber towards Laurentum. The city prepared for attack. Then it was that in the battle which raged to and fro Camilla the Volscian Amazon was distinguished for her deeds but was herself slain. When she fell the Latins were routed and tried to rush the gates of Laurentum. Turnus and Aeneas met face to face as night was falling.

THE STORY OF PALLAS

(Aeneid XII, 930–52)

Book XII is taken up with the story of the single combat between Aeneas and Turnus which was to end the long struggle between Latins and Trojans. When at last they met, Turnus' sword was shattered on Aeneas' divine armour and he turned to flee. At last he seized a huge boulder and tried to fling it at Aeneas, but his strength failed him. In answer Aeneas hurled his spear like a thunderbolt, which struck deep into Turnus' thigh.

930–52. Turnus begged for mercy, stretching out his right hand in the act of surrender. He asked Aeneas to take pity on his father Daunus and to return his body to his own people: he relinquished all claim to Lavinia as his bride. At first Aeneas was minded to spare his life, but his eyes caught sight of Pallas' belt which Turnus had proudly worn on his own shoulders. With mounting anger Aeneas thrust his avenging sword into his enemy's breast, and Turnus' life fled away.

Ille humilis supplexque oculos dextramque precantem 930
protendens 'equidem merui nec deprecor' inquit;
'utere sorte tua. miseri te si qua parentis
tangere cura potest, oro (fuit et tibi talis
Anchises genitor) Dauni miserere senectae
et me, seu corpus spoliatum lumine mavis, 935
redde meis. vicisti et victum tendere palmas
Ausonii videre; tua est Lavinia coniunx,
ulterius ne tende odiis.' stetit acer in armis
Aeneas volvens oculos dextramque repressit;
et iam iamque magis cunctantem flectere sermo 940
coeperat, infelix umero cum apparuit alto
balteus et notis fulserunt cingula bullis

Pallantis pueri, victum quem vulnere Turnus
straverat atque umeris inimicum insigne gerebat.
945 ille, oculis postquam saevi monimenta doloris
exuviasque hausit, furiis accensus et ira
terribilis: 'tune hinc spoliis indute meorum
eripiare mihi? Pallas te hoc vulnere, Pallas
immolat et poenam scelerato ex sanguine sumit.'
950 hoc dicens ferrum adverso sub pectore condit
fervidus. ast illi solvuntur frigore membra
vitaque cum gemitu fugit indignata sub umbras.

NOTES

An asterisk (*) indicates that a Note on background is also given.

Aeneid VIII, 26–183, 280–369, 454–607

26 **terras...per omnes**=*per omnes terras*. It is not unusual, especially in poetry, for a preposition to be placed inside the phrase to which it belongs. *Omnes* agrees with *terras*. In poetry adjectives are often separated by several other words from the nouns with which they agree. It is necessary therefore to observe, and make sure of, agreements most carefully before translating.

27 **alituum** - Lengthened form of *alitum*, genitive plural of *ales*. *alituum pecudumque*=*alituum et pecudum*; -*que* comes in meaning in front of the word to which it is attached as a suffix.

habebat - 'held (in its embrace)'.

28–30 **cum...procubuit...dedit** - Although *cum* meaning 'when' is regularly followed by the subjunctive in past sequence, the indicative is regularly used when *cum* is inverted, as here. In this construction the important event is expressed in the *cum* clause, and the attendant conditions such as the time, the season, the weather, etc., are given in the main clause. The *cum* clause is said to be 'inverted' because it is usual for the main event in a sentence to be found in the main clause. In this passage we might have expected, according to the more familiar practice, *cum sopor...haberet, pater Aeneas...procuhuit.*

28 **pater** - With *Aeneas* in the next line. This title is given to him both as a leader of the Trojans, and as the founder of the Roman race.

gelidi...axe - Translate in this order: *sub axe gelidi aetheris*. The genitive almost invariably is put before the noun with which it is associated. For the place of *sub*, see note on 26 above.

axe - The sky above Aeneas' head; 'vault' or 'canopy'. The phrase is a poetic way of saying 'in the open air'.

33

NOTES (VIII, 28–36)

gelidi - Not 'freezing' or even 'cold', but 'cool'. Even during the hottest months of the year, the air in the neighbourhood of Rome becomes cool in the evening. The events of this book take place in August, second only to July in Italy for heat.

29 **pectora** - Accusative of respect after *turbatus*, 'as to his heart', i.e. 'in his heart'.

bello - Aeneas was saddened at the thought of the war which had just broken out between the Trojans and the native Latins.

30 **seram** - Aeneas' rest was 'late' because his anxious thoughts had kept him awake.

31–3 Translate in this order: *huic deus ipse loci fluvio amoeno Tiberinus populeas inter frondes senior se attollere visus*. Supply *est* with *visus*.

31 **huic** - Dative of the agent after *visus* (*est*) in 33: see note on 38 below: refers to Aeneas. It is often necessary in poetry to supply some part of *sum* when translating.

ipse - The god showed himself *in person* to Aeneas, thereby doing him an especial honour.

Tiberinus - In opposition to *deus*.*

32 **pōpuleas** - From the adjective *pŏpuleus* describing the trees on the river bank.

inter - For the position of this preposition, see note on 26.

senior - Comparative of *senex* often used to mean 'aged', without any stress on the comparative meaning.

attollere - Infinitive depending on *visus* in the next line.

33–4 'him fine linen veiled with a grey-green cloak'. Tiberinus wore a linen cloak.

glauco...amictu - Instrumental ablative.

34 **harundo** - He wore a wreath of reeds around his head.

35 **adfari, demere** - Infinitives with *visus* (*est*) in 33.

curas - Aeneas' anxiety about the coming war.

36 **sate** - Vocative of *satus*, literally 'sown', from *sero*, hence *son*: addressed to Aeneas.

gente - Ablative of source or origin, with *sate*.

deum - Old form of the genitive plural (*deorum*) commonly used in poetry. Aeneas is called the 'son of the gods' because, although his father Anchises was a mortal, his mother was the goddess Venus.

Troianam...urbem - 'the Trojan city', i.e. 'the city of

Troy'. Adjectival forms of place-names are commonly used in Latin, and especially in poetry.

37 **revehis** - Aeneas could be said to be restoring Troy to Italy for two reasons. In the first place his ancestor Dardanus had originally come from Italy, and secondly, as soon as Aeneas landed on Italian soil, he built a camp at the mouth of the Tiber for his Trojan followers. Vergil regarded this as the rebuilding of Troy in Italy.*

nobis - The ethic dative used for a person closely interested in an action. Compare the Biblical 'Take us the foxes, the little foxes that spoil our vines', and the Elizabethan 'Knock me on the door and thump me'.

aeterna is in the 'proleptic' position, that is, the meaning is the outcome of the action of the verb. Hence *aeterna... servas* implies, 'keep it so that it is safe for ever'.

Pergama - The citadel of Troy, the high, defended hill around which the rest of the city was built. With it can be compared the Acropolis at Athens and the Capitolium at Rome.

38 **exspectate** - Vocative case of the perfect passive participle, agreeing with Aeneas; cf. *sate* in 36.

solo...Latinis - Dative case with *exspectate*. This usage, more frequently found in poetry than in prose, expresses the agent: it is an extension of the 'ethic' dative (see note on 37 above), in this instance showing a person or persons so closely interested in an action as to be the doer of it. These words refer to the coastal district of Latium where Aeneas has just landed. Vergil applies the local adjective *Laurens* to the whole stretch of country between the lower reaches of the Tiber and Antium (modern Anzio).

39 Supply *est* with *domus* and *sunt* with *penates*.

tibi - Possessive dative. Tiberinus promises Aeneas what he has longed for during all his wanderings, a settled home.*

ne absiste - *Ne* is frequently used with the imperative in poetry to express prohibition: the equivalent in prose is *noli* with the infinitive.

penates were the gods of the house, hence the name is often used for the house itself. Names of gods are often transferred to that over which they preside: cf. *Mars*, 'warfare', *Bacchus*, 'wine', *Ceres*, 'corn'. Aeneas brings with him the *penates* of Troy, to find a settled home for them in Italy.

NOTES (VIII, 40–1)

40 **neu** represents *et* and *ne* combined: *ne* is the negative with the imperative, cf. *ne absiste* in the previous line, and note *terrere*, singular imperative passive.

40–1 Until this time the goddess Juno has pursued Aeneas and the Trojans with relentless hatred, inflicting disasters and frustration upon them during their wanderings in the Mediterranean. She has done her utmost to prevent their reaching the land promised to them by fate. Among several reasons for her anger, the most compelling was the so-called 'Judgment of Paris'. When Paris, son of Priam, King of Troy, was a shepherd pasturing his flocks on the slopes of Mount Ida, the mountain which overlooked Troy, Juno came to him accompanied by two other goddesses, Minerva and Venus, to ask him to settle a dispute. Each of them claimed to be the fairest and Paris was asked to be their judge. Paris chose Venus, and in so doing caused deep offence to Juno. Unable to forget this insult to her own beauty, Juno loathed all Trojans and even tried to injure Aeneas and his followers though in themselves they were quite blameless. Tiberinus reassures Aeneas however by telling him that by now Juno no longer wishes to hurt him and that he will be able to settle in Italy in peace.

41 **concessere** - Alternative form of the perfect active, third person plural (*concesserunt*), very common in poetry. It is to be distinguished from the infinitive by the presence of the perfect stem, in this instance *-cess*.

deum - See note on 36.

An example of a short line, of which many occur in the poem. Some may have been left to be completed in the revision which Vergil contemplated but did not live to finish. Others seem to be intentional, and so have the effect of breaking the flow of the hexameter by introducing a dramatic pause, or emphasizing an emotional or tense moment in the narrative. This line (41) may have been kept short purposely to heighten the feeling of relief which comes to Aeneas on hearing that he is free at last from Juno's ill-will. Tiberinus pauses after making an announcement which marks a turning-point in Aeneas' fortunes so that the effect may be felt. Then he goes on to tell Aeneas of quite another matter, the prophetic white sow which he is shortly to see on the river bank (42 ff.).

NOTES (VIII, 42–5)

42 **iamque** - From *iam* and *-que*.

tibi - Ethic dative, closely associated with *inventa* (43), *iacebit* (44); see note on 38 above. You shall find', *ne... somnum*, 'so that you should not think that sleep is inventing these empty (words of mine)'; i.e. 'in case you think that this is only a dream'. Tiberinus means to say that Aeneas will find that he has not been merely dreaming when the prophecy which he is about to tell him comes true.

ne...putes - Negative purpose clause.

43 Translate in this order: *sub ilicibus litoreis sus ingens inventa*.

litoreis describes the trees which grow on the river bank, although this word is more usually applied to the sea-shore.

ilicibus - The *ilex*, or evergreen oak, is one of the commonest trees of Italy, so much so that it could hardly be omitted from any description of the countryside.

sus - The most emphatic words are often found at the end of the hexameter line. The position of *sus* serves to increase the startling effect of Tiberinus' prophecy. Interest is aroused, and held in suspense until the last word completes the meaning of the line. Few hexameter lines end in a monosyllable as here, but wherever this occurs, it is done to produce an effect. The implication is that an animal least to be expected will be a sign to Aeneas of the future greatness of Alba.*

44 **triginta capitum fetus** - 'young [i.e. a litter] of thirty head': descriptive genitive. *Caput* is often used for an individual person or animal, just as we speak of so many 'head of cattle'. The miraculous size of the litter will be of special significance, pointing to the future greatness of Alba, and later of Rome.

enixa, from *enitor*, agrees with *sus* in the line before. The use of the accusative with this verb is rare.

45 **alba...albi** - The most emphatic positions in a hexameter line are the first and the last. The position of this adjective at the beginning of the line is almost startling, and serves to stress the overriding significance of the colour of the sow. She, being white (*alba*), will give her name to the city Alba of which Ascanius is destined to be the founder (48). Her little ones too are *albi*, and signify the thirty years before the founding.

solo - Ablative of place without a preposition, a common usage in the *Aeneid*.

NOTES (VIII, 46–9)

46 This line is repeated from *Aen.* III, 393 in a passage where the same prophecy is given to Aeneas by Helenus at Buthrotum in Greece. As it is spoken by Helenus the line refers without difficulty to Italy in general. In *Aen.* VIII, however, the word *hic* can only mean the exact spot where the sow will be found, that is, on the Tiber's bank, and that is not where Aeneas afterwards built Lavinium. For this reason scholars consider that this line cannot belong to *Aen.* VIII as well as to *Aen.* III.

47 **ex quo** refers to the sight of the sow and her litter; *prodigio* is probably to be supplied. Then it can be translated, 'in accordance with which (sign)', i.e. 'this'. *quo* illustrates the use of the relative as a demonstrative pronoun, to show the connection between one sentence and that which precedes it.
annis - Ablative of 'time in which'.

47–8 Translate in this order: *ex quo ter denis redeuntibus annis Ascanius urbem Albam clari cognominis condet*. Notice how in line 48 *Albam* is put in the emphatic place at the end (note on 45): the name is, as it were, 'suspended' until the last to heighten expectation.
urbem…Albam. In Latin names of towns, cities, etc. are put in apposition (i.e. in the same case) to the words which describe them. In English 'of' is used. Translate therefore 'the city of Alba'.

48 **Ascanius**, as the son of Aeneas, represents the link between his father and the Rome which is to be. It was from Alba, the city which Ascanius is destined to found, that Romulus after many years came to found Rome.
clari…cognominis - There is a double meaning in *clari*. Vergil connects the name *Alba* with *albus*, 'white', and for this reason calls her name 'bright', but it will in the future be also 'bright' or 'famous' because of the greatness of the city and of her destiny to be the mother-city of Rome. *Alba* will take her name from the white sow, *sus Alba*; hence *clari* in one sense represents a play upon words. Vergil often plays upon meanings of names in this way.

49 **haud incerta** - 'not unsure', implying 'very sure', an example of *litotes*, a figure of speech in which a weak negative statement is made to imply a strong assertion.
incerta - Neuter plural; an example of an internal accusative which limits the action of the verb, and is found usually

NOTES (VIII, 49–53)

with intransitive verbs which normally cannot have an object.

cano - Used to mean 'foretell' or 'prophesy', because oracles and other kinds of prophetic utterances were often chanted. Tiberinus' words are awesome and are those of a god.

49–50 'Now by what means you may disentangle that which presses on (i.e. troubles) you, as victor (i.e. triumphing over your trouble), in a few words (listen) I shall instruct you.' Most of Aeneas' troubles are past now, but there yet remains the coming struggle with the Latins, the native inhabitants of Latium. Tiberinus goes on to tell him how he may obtain the help of an ally.

50 **expedias** - Subjunctive in an indirect question depending on *docebo* at the end of the line.

victor - '(you, as) victor'. Tiberinus means that Aeneas will in the end be victorious over his difficulties.

paucis - With *verbis* understood: 'in a few words', or 'briefly'.

51–4 The Arcadians were thought to have come from Greece, from a city called Pallanteum, founded by Pallas, the great-grandfather of the young Pallas, after whom he was called. Vergil connects the name of this city with Palatium, one of the highest parts of the Palatine Hill, and with that of the hill itself, the *mons Palatinus*, the most important of the seven hills of Rome. In this way Vergil accepts a legend current in his day that Greeks from Arcadia once settled on this hill long before the founding of Rome.

his oris - Ablative of place without a preposition; see note on 45 above.

genus - In apposition to *Arcades*.

52 **comites** - '(as) his comrades'; in apposition to *qui*, the antecedent to which is *Arcades* (51).

signa - The standards of King Evander. They followed him both as their king and their general.

53 **delegere...posuere** - Alternative forms of the perfect as the stems *leg-* and *pos-* show. See note on 41.

montibus - *Montes* does not always mean 'mountains' in Vergil, but often 'heights' or merely 'high ground'. The Palatine Hill is a plateau with a rocky, cliff-like edge overlooking the Tiber, about two miles in circuit.

NOTES (VIII, 54–60)

54 **Pallanteum** - Vergil gives this name to Evander's settlement on the hill where, later, Rome was to be founded, clearly suggesting that the name of the Palatine Hill (*mons Palatinus*) was derived from it. Notice *Pallantis* placed at the beginning, and *Pallanteum* at the end of the line, to show the great significance of the names.*

55 Evander is actually at war with the Latin race at the time of Aeneas' arrival, as an ally of the Etruscans, a race who lived in the country on the other side of the Tiber from Rome. They have recently driven out the tyrant Mezentius because of his cruelty. He has escaped from them and taken refuge in the country south of the Tiber, with Turnus and his people the tribe of the *Rutuli*. In obedience to an oracle, which demanded a foreign leader in their war against Mezentius, the Etruscans have invited Evander to be at their head. Too old to undertake such a task, he has promised his help instead, and even now is at war with the *Rutuli* and is being hard pressed. (*Aen.* VII, 474–82, 492–509.)

56 **socios** - 'as your comrades', i.e. the Arcadians.

57 **recto flumine** - 'straight and along my stream'. The Tiber describes many meanders between Rome and the sea, so that this is not literally 'straight and', but 'following the course of the river'. In this way Aeneas will be guided to Pallanteum; the distance would be about fifteen miles in reality.

58 Translate in this order: *ut subvectus remis adversum amnem superes*, 'so that, carried upstream by your oars, you may overcome the current (which flows) against (you)'. Tiberinus promises that the flow of the river will not hinder Aeneas' boat, but that he will make the rowing easy. The Tiber has a strong current even where it flows over the level plain below Rome.*

59 **nate dea** - Addressed to Aeneas. *nate* is the vocative of *natus*, 'born of', hence 'son of'. *dea* is in the ablative of source or origin. Aeneas was the son of the goddess Venus, although his father, Anchises, was a mortal.

primis...astris - Ablative absolute, 'as soon as the stars begin to set', i.e. 'when the dawn breaks'.

60 **Iunoni** - See note on 41. It is fitting that Aeneas should pray to Juno who is now reconciled to the Trojans.

fer - Imperative of *fero*.

NOTES (VIII, 60–6)

iramque minasque - *-que* repeated is the same in meaning as *et...et*.

60–2 Tiberinus assures Aeneas that it is Juno who now should receive his prayers, and that he will wait for those which are due to himself when Aeneas has fulfilled his appointed tasks, as *victor* implies.

62 **ego** - Tiberinus is both the personification of the river, and the river Tiber itself: he talks of himself therefore in terms of the river. Up to this point he has not yet revealed his identity to Aeneas.

64 **caeruleus** - The water of the Tiber is of a murky greenish-yellow colour which often appears quite green, due to the vast quantities of silt which are carried down from the upper course. *caeruleus* (from *caelum*) means 'blue' or 'grey', the colour of the sky: neither meaning describes the Tiber, but the word is often used to describe gods connected with water, as well as water itself. Vergil points out the derivation of the epithet in the words *caelo gratissimus amnis*, and at the same time implies a deeper meaning than that of mere description, the thought that the Tiber is 'dear to heaven', i.e. 'loved by the gods'. *caeruleus* in this passage is an example of a 'stock epithet', that is, an adjective so commonly attached to a person or thing as to be used regardless of whether the description is really fitting in any particular instance. Although water of the Tiber is not *caeruleus*, water in general can carry this epithet.

Thybris - Another older form, probably Etruscan, of the name *Tiberis*, the usual Latin form for 'Tiber'. Vergil uses this name to give a feeling of antiquity to line 65. 'Here is my great home; it issues forth (giving) life to lofty cities.' Aeneas was asleep on the river bank near the mouth when Tiberinus appeared to him; there is no doubt therefore that *exit* refers to the end of the Tiber's course. *caput* = 'source of life'.

65 **urbibus** contains a prophecy of the future Rome to be built on the seven hills skirted by the Tiber, and also refers to the many ancient cities which stood on high hills and eminences along the upper course of the river.

66 **fluvius** - Tiberinus, who descends to his home in the depths of the stream; Vergil perhaps imagined him as living in some grotto deep under the water; see *ima petens* in the next line.

NOTES (VIII, 68–77)

68 **orientia** - Present participle of *orior*, agreeing with *lumina* in next line.

68–70 Aeneas observes Roman customs when he turns to the east, and takes up water in his cupped hands as an act of purification before prayer. It was customary too to stand with the hands uplifted and palms turned upwards.

70 **aethera** - Greek form of the accusative singular used in a word which is Greek in origin. This form is commonly found in the *Aeneid* with words borrowed from the Greek as well as in Greek proper names.

71 **Laurentes nymphae** - Laurens is a local adjective which can be applied to all, or any part, of the coastal district between the Tiber and Ardea to the south, and to the inhabitants themselves who are sometimes called *Laurentes*. Aeneas prays to the spirits of all the rivers and streams which flow across that country which is now called the 'Roman Campagna'.

amnibus - Dative of possession.

unde refers to *nymphae*. Being goddesses of springs they are thought of as giving birth to streams and rivers.

72 **Thybri** - Vocative case.

genitor - Rivers were often called 'father'; this was both a title of respect, and also natural because rivers give life to the country through which they flow.

73 **arcete periclis** - An example of *hypallage*, when the usual relations of persons or things are changed over. Aeneas prays 'ward me off from dangers', but the normal expression would be 'ward dangers off from me'.

74–5 'in whatever spring a pool keeps you, taking pity on our misfortunes, and from whatever ground you gush out most beautiful...'.

74 **quo...cumque** = *quocumque*, an example of tmesis, the cutting up of a word: ablative of *quicumque*, agreeing with *fonte* in the next line.

lacus - The original meaning is anything hollow, hence 'that which lies in a hollow', i.e. a lake or a pool; cf. 66 and note.

76 **celebrabere** - Alternative form of *celebraberis*, found fairly frequently in poetry.

77 **corniger** - Vocative, agreeing with *fluvius*. Rivers when personified were often thought of as horned, because horns

42

NOTES (VIII, 77–83)

were considered to be a sign of strength. Sometimes, in sculpture, rivers were represented as bulls.

Hesperidum - 'belonging to the West', in other words, 'to Italy'. The Tiber is addressed as the lord of all the rivers of Italy. It is, in fact, the largest river of west central Italy and receives as its tributaries most of the streams which drain that part of the country.

fluvius - Take closely with *celebrabere* (76) ('as...').

78 'Only be present, and more closely (at my side) strengthen your divine powers.' Aeneas prays that Tiberinus will speedily carry out what are yet only promises, and in doing so prove the truth of his words.

adsis...firmes - The jussive subjunctive used regularly in prayers.

79 **geminas** - Aeneas promptly equips two boats for his journey up the Tiber to Pallanteum.

80 **remigio** - 'with oars'.

81 **oculis** - Dative, closely associated with *mirabile monstrum*, in apposition to *sus* (83). The sow, found so soon after Tiberinus' promise to Aeneas, is regarded as an 'omen' or 'portent' because with the miraculous litter of thirty young it is prophetically symbolic of the greatness and prosperity of the city of Alba, which in due time Ascanius will found, and which in turn will be the mother-city of Rome. The appearance of the sow marks one of the most important events in the story of the *Aeneid* because it serves as a link between Aeneas and the Trojans and future Rome. The word *monstrum* usually implies a sign from heaven of misfortune to come, but in this passage the very opposite is meant.

82 **per silvam** - They catch sight of the sow gleaming white among the trees which grow on the river bank. By contrast the white colour (*albus*), so essential to the omen, is emphasized against the colour of the trees. This is the colour after which future *Alba* will be called (42–8).

83 **viridique in litore** - 'and on the grassy river-bank'. *litus* is usually applied to the sea-coast, but frequently in poetry to the bank of a river.

sus - The position of this monosyllabic word at the end of the line gives an impression of excitement and astonishment (43 and note). Notice how the order of words in the preceding lines from 81 has led up to this last word, and has kept the reader in suspense. It is especially important to keep to this order in translation to give the same effect.

84–5 In obedience to Tiberinus' command (60–1) Aeneas sacrifices the sow and all her young to Juno so that the goddess may be favourable to the Trojans. Up to this time she has pursued them with unchanging hatred (41 and note), but is now ready to put aside her anger against all Trojans. This offering will finally placate her. We must imagine Aeneas and his men building an altar of sods (85) and being delayed for several hours while they carry out the sacrifice. They are not ready to set out on their journey until nightfall (86).

84 **quam** refers to *sus* in the previous line. This relative is used with a demonstrative meaning; translate as 'her'.

pius - Not the equivalent of the English word 'pious', but 'dutiful', 'affectionate', or 'god-fearing'. It is especially used of one who is loyal to parents, country, or the gods, and is so often attached to Aeneas as to be almost a title. He is *pius* especially because he rescued his aged father Anchises from burning Troy by carrying him on his shoulders to safety (*Aen.* II, 717–22). He is also to be the founder of the Roman race and in this respect faithfully carries out the tasks entrusted to him by the gods. He is the very pattern of a chosen, dutiful servant.

tibi enim, tibi - The sudden change to the second person is as striking as Juno's change-over from hatred to approval. The words carry a joyful note.

enim - Used in a corroborative sense, 'to you, yes, to you'.

85 **sacra** - The vessels and instruments used in sacrifice, such as platters, bowls, jugs, wreaths, sacrificial axe and knife.

cum...aram - *sistit*, which is the act of leading the animals to the altar, should in reality come before *mactat* which is the word for the slaying of the victims: this is an example of *hysteron proteron*, a figure of speech in which the logical order is inverted: cf. in English 'to put the cart before the horse'. Aeneas first brings the sow to the altar and then performs the sacrifice.

86 At this moment Aeneas and his men embark and begin the voyage up the Tiber. They row all night long, as is fitting,

NOTES (VIII, 86–93)

since this is in August (94 and note, below), one of the hottest months in Italy, second only to July.

86–7 Translate in this order: *Thybris ea nocte, quam longa est, fluvium tumentem leniit.* Although the Tiber flows over level country below Rome until the sea is reached, the current is swift and powerful. To row against it would be a strenuous undertaking involving hard rowing.*

quam longa est - Literally 'in that night as much as it was long', i.e. 'all night through'.

refluens - The river did not actually 'flow backwards', but stood still so that the current was held back, and the water became like a standing pool as the next line shows. This stilling of the stream is a special favour given to Aeneas by Tiberinus and the fulfilment of his promise that Aeneas will be able to overcome the difficult journey upstream (58).*

88 **mitis** agrees with *Thybris* (86).

in morem - With the genitive (*paludis*), 'after the manner of...', or 'like a...'.

paludis - Usually describes a marsh, but here a stretch of deep water.

89 **sterneret** - Consecutive subjunctive after *ut* in the preceding line, 'so that he laid a smooth surface on the waters'; cf. *abesset*. The result is described which followed when the flow of the current was stopped.

remo - Dative of advantage.

90 **iter...celerant** - 'they set out on their journey and speed (on their way)'.

rumore secundo - There are several explanations possible for this expression. It could describe the cries of the rowmaster as he called to the rowers, or perhaps the chanting of the rowers as they got well under way, or even the splash of the oars and the rippling of the water against the sides of the boats. Whichever is the correct interpretation, there is at any rate a suggestion, in the smooth sounds of the words, of the ease with which they move along.

91 **uncta** - Keels were often greased to enable them to slip easily through the water (*vadis*).

abies - The wood of which the ships were made, 'the hulls of firwood'.

91–3 'wondered too the waves, wondered the grove unaccustomed (to the sight) that gleaming afar warriors' shields in the stream were floating, and painted ships (as well)'. These contain an example of the 'pathetic fallacy', a turn of expression common in poetry and familiar in Vergil, by which human feelings and emotions are given to objects of nature; cf. the Biblical passage, 'The mountains and the hills shall break forth before you into singing, and all the trees of the field shall clap their hands' (Isaiah lv. 12). Notice the repetition of *mirantur*, *miratur*, which heightens the effect of astonishment.

93 **scuta** - These were hung along the ships' sides and would be reflected in the water.

virum - Old form of the genitive plural (*virorum*): cf. 36 above and note on *deum*.

fluvio - Ablative; see 45 above and note on *solo*.

innare - The accusative and infinitive construction is used after *mirantur* (91) and *miratur* (92).

94 **olli** - Archaic form of *illi*. Vergil frequently makes use of old-fashioned forms and words to give an effect of the dignity of bygone days and of the greatness which is attached to antiquity. This is an especial mark of epic poetry.

fatigant - Literally, 'they wear out', but here the meaning is not much more than 'spend' or 'keep on their way', though the word by an inversion of meaning seems to hint at the work of rowing. The distance, taking into account the winding course of the Tiber over which they must row, is about eighteen miles. This does not take them all day as well as all night, for they reach the site of future Rome at midday (97): *diem* therefore means part of the next day.

95 **flexus** - This is a true description of the Tiber's course below Rome; there are many meanders, some of them very considerable, which add to the mileage which the Trojans have to cover.

teguntur - They keep under the shadow of the trees which grow along the banks to shelter from the August heat.

96 **secant** - 'they pass between' the woods on either bank. These apparently overhang the water, but are not over-arching because the Tiber is too wide for trees to meet overhead.

97 It is noon when they come within sight of Pallanteum.

medium caeli - 'midway in the sky'. The Romans believed

that the sun moved over the earth (which they thought was flat) during the day, and crossed from west to east underneath it during night.

98 **arcem** - The *arx* was the defended part of an ancient city, and therefore was usually a hill, 'the stronghold' which served as a refuge in times of danger. Here the Palatine Hill is meant, one of the most important of the seven hills of Rome. Where this hill fronts the Tiber, its sides are notably cliff-like and precipitous and certainly look impregnable. The legends concerning the founding of Rome told how the first settlement was made on this hill so that it became the ancient heart of the City.*

98–100 These lines are especially striking because in them is related the first glimpse which Aeneas and his men have of the place where Rome of the future will be built. All unknowing the Trojan founder of the Roman race looks at a scene one day to be the most sacred of all to all true Romans.

98–9 **rara...vident** - The Trojans saw a hut settlement clustered on the brow of the Palatine Hill. From primitive times the Romans lived in huts made of wattle and reed with pointed roofs. These persisted in the country even after houses were built, and examples can still be seen even today on the Roman Campagna, inhabited by shepherds.

99–100 A contrast is drawn between the poor huts of Evander's settlement and the splendid buildings which were to be seen in Vergil's own day. By then the hill had become a favoured residential quarter of Rome where rich men of distinction had built themselves villas. In later times the Palatine Hill became completely covered with the palaces and gardens of emperors. The English word 'palace' is derived from the name of the hill.

99 **nunc** - Vergil speaks of his own time, drawing attention to the appearance of the hill as he and his contemporaries know it: the word contrasts with *tum* in the next line.*

100 **res inopes** - '(his) poor estate'.
Euandrus - An alternative, but rare, form of *Euander*.

101 **ocius** - They row all the more eagerly, encouraged by the joyful sight of Evander's huts.
urbi - Not a city in the full sense, but 'settlement' or even 'village' in this context. There is something prophetic, however, in the use of *urbi* in connection with the Palatine

NOTES (VIII, 101–6)

Hill. The dative is as regularly used with *propinquo* as is *ad* with the accusative, although it is a verb expressing 'motion towards'.

102 **Forte** - Not 'by chance' in this context, but 'as it happened'.

die...illo - 13 August (106 and note).

sollemnem - The adjective *sollemnis* carries the root meaning of 'yearly' or 'annual', being a compound of *sollus* 'whole', and *annus*. It is used especially of religious rites and from that has the derived meanings of 'established', 'appointed', and hence by an easy transition that which the English word has inherited, of 'solemn'. In this context the word is seen in its most primitive meaning, but at the same time the feeling of reverence attached to a great yearly festival is there in full.

rex Arcas - Evander himself.*

103 **Amphitryoniadae** - 'The child of Amphitryon', i.e. Hercules. He was not the true son of Amphitryon but of Zeus, who visited Alcmena his mother while her husband Amphitryon was absent from home on military service. Thus Hercules was worshipped as a semi-divine hero. The ending *-ades*, added as a suffix to a name, means 'son of' or 'relation of'. The form is called a 'patronymic': cf. *Aeneades*, 'son of Aeneas', i.e. *Ascanius*. The placing of this very striking name, made up of seven resounding syllables, at the beginning of the line summons up a sense of awe. We feel as the Trojans felt at the realization that they had arrived at a supreme moment in the rites of the hero-god.*

divis - It was customary to include most of, if not all, the gods in sacrificial prayers, although the sacrifice might be offered to one alone.

104 **ante urbem** - The place of sacrifice was on the lower ground along the bank of the Tiber, at the foot of the hill on which Evander's huts were to be seen.

in luco - Temples and altars were usually inside a sacred grove.

Pallas - The first sight we have of Pallas is as he takes part in the sacrificial rites at his father's side (*una*).

huic - Possessive dative referring to Evander.

una - Adverb, 'together'.

105 **senatus** - In contrast to *iuvenum*. The line implies that both young men and old were present.

106 **tura** - From *tus*. This was burnt on the altars.

cruor - A heifer was killed in Hercules' honour on the evening of the day sacred to him, 12 August, and on the next day the worshippers took their part in the rites by eating the roasted flesh at a communal meal. Aeneas arrives on 13 August at one of the most solemn moments in the festival, when the flesh, *cruor*, is ready to be handed round (180 below). It is interesting to see how Vergil thought that the worship of Hercules was older than the founding of Rome, and that it was inherited from Evander by the Romans.*

107 **ut**, with the indicative (*videre*), means 'when' in a purely temporal sense.

videre - Alternative form of the perfect (41 above and note). The Arcadians are the subject.

rates - The two boats (*biremes*, 79–80 above) in which the Trojans had rowed up the Tiber. They are described as high because, like all Roman ships of this kind, they would have a high poop perhaps decorated with a figure-head. This, standing high, would cause them to be easily seen as they approached the grove. The oldest quay in Rome along the Tiber's bank was at the place where the Trojans landed.

108 **adlabi...incumbere** - Infinitives depending on *videre* in the line before, in the accusative and infinitive construction.

nemus - Cf. *luco* (104). The worshippers catch sight of the ships' prows through the trees.

tacitis - A 'transferred epithet'. This word really describes the oarsmen, but is attached grammatically to their oars. They move silently now as they draw in to the landing-place and let the boats slide (*adlabi*). They are quiet too in honour of the sacrifice which they now see to be taking place.

109 **terrentur** - The Arcadians are panic-stricken not merely by the arrival of strangers, but by the interruption to the rites of sacrifice. In Roman superstition, if any unlucky incident intruded on religious observances, these became nullified and had to be observed all over again from the very beginning. If this were not done, they believed that misfortune was sure to follow as a result of such an insult to the gods whose sacrifice had been contaminated.

109–10 **relictis...mensis** - Ablative absolute. In poetry words which go closely together are frequently separated and may even be placed in different lines.

110 **mensis** - Either the tables on which the sacrificial flesh has

been laid as an offering to Hercules, or the sacrificial banquet.

audax - The first hint of Pallas' doom. He takes the lead in this crisis and then goes alone to meet Aeneas. Vergil often uses this word to imply brave without ultimate success.

quos - Relative used as a demonstrative.

110–11 Pallas calls the worshippers back to the feast to prevent its being interrupted and the ceremony spoilt.

112 **tumulo** - The river bank.

113 **ignotas** - Recognizing the Trojans as foreigners, Pallas realizes that they have not come this way before: it is therefore 'unknown' to them.

114 **qui genus?** - *qui*, masculine plural of the interrogative pronoun *quis*; *genus*, accusative of respect, 'who (are you) as to race?', i.e. 'what is your race?'

unde domo? - 'from where, from home?', a condensed expression for 'where have you come from, and where is your home?'

pacemne=*pacem* and the suffix -*ne* used in a direct double question and followed by *an*.

115 **puppi...ab alta** - *ab alta puppi*: *puppi* is the alternative form of the ablative singular.

fatur - From *for*.

116 **olivae** - The olive has always been associated with peace. An olive bough hung with strands of wool (*vittae*) (128 below) and carried in the hand was a sign that the bearer came with peaceful intentions.

117 **Troiugenas** - 'men born in Troy', a far stronger word than *Troiani*. Notice the emphatic place of this name at the beginning of Aeneas' reply. He is quick to announce their race to show that they are not enemies.

Latinis - Evander has already promised his help to the Etruscans in their struggle against the *Rutuli*, a Latin race (55 and note, above). He is all the more likely to receive the Trojans in friendship since they have enemies in common.

118 **quos** refers to *Troiugenas*.

illi refers to *Latinis*.

119 **ferte** - Imperative from *fero*.

lectos - From *lego*: agrees with *duces* in the next line.

120 **Dardaniae** - Closely associated with *duces*. Aeneas says in so many words, 'We are leaders from Troy who have come to

NOTES (VIII, 120–30)

ask your help'. *Dardania* was the oldest name for Troy, called after Dardanus the first founder (134 below), who himself had come from Italy.*

venisse - Notice the accusative and infinitive construction after *dicite* in the preceding line.

121 **percussus** - Pallas has heard of the fame of Troy and so is struck with astonishment on hearing the name.

122 **egredere** - Imperative; cf. *adloquere* in the next line.

coram may be either a preposition or adverb, but here is an adverb, 'face to face'.

123 **penatibus** - Dative case, regular with *succedo* although it is a verb expressing 'motion towards'; cf. *propinquo*, and note on 101, above.

hospes - '(as a) guest'.

124 'He received him with (outstretched) hand, and clasping his right hand clung to it.' Pallas extends to Trojan Aeneas the warmest possible sign of friendship.

125 **luco** - They enter together the grove where Evander presides over the festival.

126 **amicis** - Adjective agreeing with *dictis*.

127 **optime** - Vocative, addressed to Evander.

Graiugenum - The old form of the genitive plural, often used with names (note on 41, and cf. *Danaum*, 129 below). The name answers to, and carries the same dignity as, *Troiugenas* (117).

128 **comptos...ramos** - The olive branches which Aeneas carried were twined with strands of wool (*vittae*) as was the custom when they were carried by persons who wished to ask for peace. The olive, sacred to Pallas, the goddess of the arts, was a fitting symbol of peace.

129-30 'For myself, I was not afraid because you were a leader of the Danai, and an Arcadian, and because by your origin you were kin to the two sons of Atreus.' Aeneas excuses himself for presenting the olive branch. He is assuredly not afraid of meeting Greeks, but he wishes to be received for his own merits, and to show that he comes peaceably in obedience to the gods and the ties of kindred (131-3). The Trojans had had ample cause to hate and fear the Greeks, since they had utterly destroyed the city of Troy after a siege of nine years. Aeneas shows true bravery in seeking to meet Greek Evander, especially since he is related to the two

NOTES (VIII, 129-42)

leaders of the Greek host, Agamemnon and Menelaus (*Atridae*).

130 **fores** - Imperfect subjunctive from *sum*: a subjunctive expressing a 'rejected reason' where the sense is, 'not because (as someone may say)...but (in reality) because...'. The subjunctive shows what someone other than the speaker might say or think. This use is very close to that of the subjunctive in subordinate clauses in indirect speech (*oratio obliqua*).

131-3 Aeneas explains the motives and powers by which he has been led to seek out Evander. *virtus, oracula, patres, fama* are all the subjects of *coniunxere* and *egere*; *me* (131) is the object.

131 **mea...virtus.** When Aeneas escaped from the burning of Troy, he rescued his father Anchises, then a feeble old man, by carrying him on his shoulders (note on 84 above). This heroic action brought him fame and gave him a claim to be called brave. Aeneas is not really boasting of his bravery but rather explaining his identity. During the many years of his wanderings around the Mediterranean, the news of the fall of Troy and of his action has gone before him and become well known.

oracula - Aeneas has been told twice in prophecy about Evander, by Apollo at Cumae in *Aen.* VI, 96-7, in the words: 'The first avenue of safety, a thing which you least expect, shall be opened from a Greek city', meaning Evander's city, Pallanteum, and again by Tiberinus (49-54 above).

divum - Old genitive plural (note on 41 above).

132 'our related fathers, (and) your fame spread in (all) the earth', i.e. 'known throughout the world'. Aeneas reminds Evander (134-42) that the Trojans and Arcadians have a common ancestor.

133 **fatis egere volentem** - 'and the fates drove (me) willing', i.e. 'brought me here the willing servant of destiny'. Aeneas has been obliged to come to seek Evander's help, but at the same time he comes gladly because he knows that he is ordained by the gods, whose will is expressed by fate (*fatis* in 133), to found a new race in Italy. He is only too ready to obey the dictates of his grand destiny.

volentem agrees with *me* understood.

134-42 Far from fearing to meet Evander, Aeneas tells him how they both have the same ancestor and so are kindred. Aeneas

NOTES (VIII, 134–8)

claims as his forbear Dardanus, the founder of the city of Troy; Evander, on the other hand, had for his father Mercurius. Dardanus and Mercurius were cousins, being sons respectively of Electra and Maia, daughters of Atlas, who was in this way their common ancestor. The following table shows their relationship:

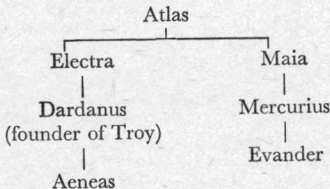

135 **Electra** - Ablative of source, with *cretus*.
Atlantide agrees with Electra: notice the patronymic, 'daughter of'. This form is not often found in the feminine.

136 **advehitur** - Passive of *adveho*, literally 'I am carried', hence 'I ride' or even 'sail': in this context translate 'journeyed'.
Teucros - Accusative of 'motion towards'; the preposition *ad* is omitted, as with the names of towns. The name of the people, as often, is used for the country; translate 'to (the land of) the Trojans'. This is an example of prolepsis, a statement made too soon, because there were no *Teucri* until after Dardanus came to that country. Aeneas means to say 'the land where the *Teucri* afterwards lived'.
Atlas - A giant who was believed to hold up the sky on his shoulders. He was thought to live in north-west Africa where are the mountains which still have his name (the Atlas Mountains): by an extension his name has survived in that of the Atlantic Ocean.

138 **vobis** - Possessive dative: 'your father is...'.
Mercurius - Mercury was the messenger of the gods, and also of trade and merchants. He wore a winged hat, and winged sandals, and carried a staff twined with serpents.
quem - i.e. Mercurius.
candida - Maia is described as 'bright' or 'gleaming' because she became a star, the brightest of the seven sisters, daughters of Atlas, called the Pleiades.

139 **Cyllenae** - A mountain in Arcadia in Greece; notice the emphatic position to stress the connection of Maia with Arcadia.

conceptum...fudit refers to Mercurius: 'conceived and bore'.

140 **Maiam** - Also in an emphatic position; object of *generat* in the next line.

auditis - Dative of the perfect passive participle. 'If we put any trust in things heard', i.e. 'in what we hear'.

quicquam - Neuter singular of *quisquam*, object of *credimus*; an internal accusative (49 above and note). *Credo* regularly governs the dative, as is seen in *auditis*, but *quicquam* implies 'if we put any belief in...'.

140-1 **Atlas, idem Atlas** - Very emphatic, stressing the common source from which both Aeneas and Evander claim descent.

141 **generat** - Not the historic present, but a special usage to indicate a fact which remains unchanged. Atlas was the father of Maia, continues to be, and always will be.

143-4 Aeneas, relying on the ties of kindred, has approached Evander directly without any preliminary overtures or any effort at diplomacy.

143 **his** - Neuter plural, referring to all that Aeneas has just related.

legatos - Object of *pepigi* (144) as much as is *temptamenta* (144). The meaning of the verb ('settle' or 'fix') is not easily attached to *legatos* but is easily understood with *temptamenta*. This is an example of zeugma, in which two words are used with one verb, but only one of them really fits the meaning. This grammatical term comes from the Greek for 'a joining' or 'yoking together'.

144 **tui** - Genitive of *tu*. The use is 'objective' because *tui* stands in the same relationship to *temptamenta* as would an object to the corresponding verb *tempto*.

pepigi - From *pango*.

144-5 'me, me (i.e. my own person) myself, and my life have I offered (to you) and have come (as) a suppliant to your doors (lit. "thresholds")'. Notice the emphatic repetition. Aeneas declares that he has come *in person* to Evander, and has not, as would have been usual, sent messengers ahead.

NOTES (VIII, 145–52)

145 supplex - Nominative case in apposition to the subject of *veni*.

146–7 insequitur comes twice in meaning, but is only once expressed; this is a common usage in Latin (cf. 149 below): *gens eadem (nos insequitur)... quae te insequitur*.

146 Daunia - *Daunius* is another name for Rutulus, after Daunus, father of Turnus. See Vocabulary for other names very like this and be careful to differentiate.

147–9 'if they were to drive us (out), they believe that nothing would prevent them from putting all Hesperia utterly beneath their yoke (lit.: "they believe that there will be nothing wanting so that they may not put...") and from holding (the land) against which the sea washes above, and (that) against which the sea washes below'.

147 pellant - Subjunctive for two reasons: (1) in a subordinate clause in *oratio oliqua* after *credunt*; (2) in an 'if' clause expressing a probability.

afore - future infinitive of *absum* depending on *credunt* in the accusative (*nihil*) and infinitive construction.

148 quin - Used in a regular construction here after a negative (*nihil*) with the meaning of 'prevent', and followed by the subjunctive (*mittant, teneant* in next line).

149 adluit comes twice in meaning: supply after *supra* (cf. 146 above).

mare... supra... infra - The two seas which wash the coasts of Italy were called *mare superum* (the Adriatic on the east) and *mare inferum* or *Tyrrhenum* (the Tuscan, on the west).

150 daque - *Da*, imperative of *do*, and *-que*.

151 sunt - Supply *nobis* from the preceding line.

rebus spectata - 'tested (and approved) by circumstances'. After enduring the nine years' siege of Troy and long wanderings in the Mediterranean, the Trojans had become hardened to war and danger.

152 Dixerat - The pluperfect tense implies that Aeneas has finished what he wished to say: he makes, as it were, an end of speaking.

ille - Evander.

os oculosque loquentis - Lit. 'the mouth and eyes of him (Aeneas) speaking (i.e. as he spoke)'. Evander looked closely at Aeneas' features while he was speaking to him.

NOTES (VIII, 153–61)

153 **iamdudum** goes closely in meaning with *lustrabat*. The imperfect tense is used because the action has been continuous all the time that Aeneas has been speaking. Evander has long been scanning Aeneas' features to see if he can find any likeness to his father Anchises whom he had once met.
et connects *os oculosque* with *corpus*.
lumine - 'with his eye', a common meaning of *lumen* in Vergil, but more usually in the plural.

154 **pauca refert** - 'he speaks a few (words)', i.e. 'he replies briefly'.
ut is used with the indicative in the next line (*accipio, agnosco*) in an exclamation, meaning 'how'.
fortissime - Vocative, addressed to Aeneas.
Teucrum - Genitive plural (notes on 41 and 127 above).

155 **accipio** - Evander receives Aeneas because he recognizes his likeness to Anchises.
libens - Adjective agreeing with *ego* understood, the subject of *accipio*.

156 **Anchisae** - Genitive case; *magni* agrees with it.

157–9 **memini** governs the accusative and infinitive construction, the most important words of which are *Priamum, invisere*.

157 **visentem** - Present participle from *viso*, not to be confused with the verb *video*, agreeing with *Priamum*.

158 **Laomedontiaden** - A Greek name in which is preserved the ending *-en* of the Greek accusative: for the patronymic see 103 and note, above.
petentem agrees with *Priamum*.
Salamina - Greek accusative of a Greek name.—Arcadia was a mountain district in southern Greece.

159 **invisere** - Infinitive of *inviso* (do not mistake for *invideo*).

160 'at that time early (lit. "first") youth was clothing my cheeks with down'. Evander was only a young man when he first saw Anchises.
mihi - Possessive dative.
vestibat - Old form of the imperfect (*vestiebat*). This form is common in Vergil in fourth conjugation verbs.
iuventas - Nominative singular; the commoner form of this word is *iuventus*.

161 **duces Teucros** - Those who attended King Priam on his journey.

NOTES (VIII, 162–9)

162–3 'but taller than all moved Anchises'.
  cunctis - Ablative of comparison.
  Anchises - Notice the emphatic position at the beginning of the line.

163 mihi - Possessive dative.
  iuvenali agrees with *amore*. As a result of his boyish admiration, Evander was seized with a great desire to talk to Anchises and to take him to his home.

164 compellare, coniungere - The infinitive is used in poetry with the verb *ardeo* (163) when the meaning is 'burn to', i.e. 'long to'.
  virum - In the full sense of 'a warrior', 'a mighty man'.

165 accessi - Supply *eum*.
  cupidus - 'eagerly'. Adjective with an adverbial meaning: a common usage in poetry.
  Phenei - Pheneus was presumably Evander's chief city in Arcadia: there was a lake near by of the same name.

166 Lycias - The Lycians were a tribe of Asia Minor especially skilful in archery, hence the adjective *Lycius* is often applied to bows and arrows, etc. as a 'stock epithet'; but since Priam had come from Phrygia in Asia Minor, his gift may be in reality 'Lycian'. This would be likely all the more to delight the heart of the boy Evander.

166–7 The outline of the sentence is: *ille mihi pharetram... sagittas...chlamydemque...dedit*.

167 The spondaic ending, especially remarkable because it is made up of one word (*intertextam*), has the effect of slowing down the rhythm of the hexameter. Evander gives full expression to his delight at the gift, lingering over the memory of how it was embroidered with gold.
  discedens agrees with *ille* (i.e. *Anchises*) in the line before, '(as he was) going away', or 'on his departure'.

168 frena...bina - 'a pair of...' or 'two sets of...'.
  aurea - The position, separated at such a distance in the line from the word which it qualifies, serves, by isolating this word, to emphasize all the more Evander's delight at the lavishness of the gift. The placing of *aurea* next to *Pallas* seems also to hint of his feelings of pleasure.

169 iuncta est - The perfect tense implies a quick gesture. As Evander speaks he stretches out his right hand to Aeneas, who immediately clasps it.

mihi - Dative of the agent (note on 38 above).

dextra - The relative is *quam*.

170 **crastina** - With *lux*.

cum...reddet - Latin requires a future in a subordinate clause dependent on a main clause which also refers to the future. *reddet* is subordinate to *iuvabo* (171). The English idiom requires a present tense; translate therefore 'as soon as the daylight *returns*...'.

terris - Dative with *reddet*.

171 **laetos** - i.e. *viros*. They will be happy in their eagerness to come to the help of the Trojans.

iuvabo - Supply *te*.

172 **sacra haec** - The sacrifice to Hercules. *haec* implies a gesture (the deictic use of the pronominal adjective) as Evander points to the scene of the Festival.*

quando - An unusual and at the same time poetic use with the indicative (*venistis*) to mean 'since'.

amici - '(as) friends'. Very emphatic since strangers were not allowed to take part in the rites of Hercules, but Aeneas and his men have now been accepted as friends and even kinsmen (134-42 above).

173 **annua** agrees with *sacra* in the line before.

nefas - Supply *est*. The sacrifice cannot be postponed because it is an annual one and must take place on this particular day (note on 102 above). Evander shows especial courtesy to Aeneas when he even goes so far as to mention this.

faventes - From *faveo*, agreeing with *vos* understood from *celebrate*. This verb is especially used in connection with ritual observances for refraining from any word or action which might be unlucky. Sometimes, to avoid this, silence was expected of the worshippers, and, should any unlucky event happen, the whole sacrifice with all the detail of ritual would have to be renewed from the beginning.

174 **sociorum** - Another insistence on the friendship now established between Arcadians and Trojans (cf. 173 above). This understanding is essential if the Trojans are to be allowed to take part in the sacrifice.

175 **dicta** - Supply *sunt* and in translation 'words'.

dapes - A word used especially in ritual for the portion of the feast which was offered to the god.

sublata - See *tollo*.

NOTES (VIII, 175–81)

175–6 Translate in this order: *dapes et sublata pocula reponi iubet gramineoque sedili ipse viros locat*. The feast has been interrupted by the arrival of the Trojans, but not broken off (110–11 above). Now, recognizing their kinship and after making a bond of friendship with them, Evander orders the feast to continue.

sublata reponi pocula – 'the goblets, which had been removed, to be replaced'. These have presumably been put aside on the appearance of the Trojans as a precaution against possible contamination of the sacred rites. Now they are set in place again.

176 **ipse** – Evander conducts them *in person* as a sign of courtesy.

sedili – A word of especial significance and in an emphatic position at the end of the line. The worshippers at the festival of Hercules used to sit, thereby observing a custom older than that of reclining at meals which was the general practice among the Romans, although in this connection the ritual may be Greek.

177–8 'conspicuous (before all) on a couch and the hide of a shaggy lion he receives Aeneas, and welcomes him with a maple chair'. The couch is spread with a lion's skin. These lines also stress the custom of sitting at this feast.

177 **leonis** – It is fitting that the skin should be that of a lion, because the lion was closely associated with Hercules and the skin was one of his attributes. In statues it is usually shown draped over his shoulders.

179 **lecti iuvenes** – The young men whose duty it is to assist at the rites. Some of them may only have been boys. They are *lecti* because they belong to the two priestly families, the Potitii and Pinarii who originally had sole charge of the cult of Hercules.

arae – The great altar of Hercules, known as the *ara maxima*.*

180 **viscera** – 'flesh', either the inward parts of an animal, or all the flesh without the skin.

181 **dona...Cereris** – 'gift of well milled (lit. worked at) corn', i.e. 'cakes of good bread'. According to the Roman custom these would be flat and round. Ceres was the goddess of the corn: as is common with the names of deities, her name is transferred to that which belongs to her and is under her special tutelage: cf. *Bacchum* in the same line for 'wine'.

59

NOTES (VIII, 183–285)

183 **perpetui...bovis** - 'an oxen's chine uncut'. *perpetui* is a transferred epithet. The meaning really should be attached to *tergo*. An especially favoured and generous share was set before the Trojans.

tergo, extis - Ablatives with *vescitur* (182): *vescor* regularly governs the ablative case.

lustralibus extis - 'sacrificial entrails', of the heifer which has been offered in sacrifice. They are 'holy' because they have first been put on the altar. They are set before the Trojans as another mark of courtesy.

280 **Devexo...Olympo** - 'evening drew near the lower slopes of Olympus'.

281 **primusque Potitius** - The sacrifice is carried out by the members of two families, the Potitii and the Pinarii: the Potitii were the more important of the two, and always took the first place, hence *primus*. The Potitius mentioned in this line is regarded as the founder of the priestly family.*

282 **pellibus** - Instrumental ablative.

in morem - 'according to custom'.

flammas - This detail enlivens the scene of sacrifice. The priests carry torches in procession, and then set them up in sconces so that the feasting may be continued into the night. Pinewood was often used for torches of this kind.*

283 **instaurant** - A technical word used in connection with religious rites, meaning 'renew'. The rites have been interrupted so that Evander may tell the tale of Hercules and Cacus. Now the feasting begins again.

mensae...secundae - The second part of the feast. Probably fruit is now served, since meat had been eaten earlier in the day, but it was customary in the worship of Hercules to consume entirely the remains of the sacrificed victim after the manner of Greek ritual. The meaning may be merely that there was a second large sacrificial meal.*

284 **aras** - Offerings to the god in whose honour the sacrifice was made were placed on the altar. *aras* is a poetic plural.

285 **Salii** - Dancing priests more usually connected with Mars, the god of war. The word is derived from *salio*, 'leap'. They dance in war dress and chant. This brings a picturesque interlude but is an integral part of the ritual.*

NOTES (VIII, 285-90)

incensa altaria - The offerings to Hercules were burnt up on the altar, though this was not always the custom in Roman ritual.

286 **populeis...ramis** - The Salii wear wreaths of poplar in honour of Hercules, to whom this tree was sacred.

evincti tempora - 'having their temples bound...'. This construction made up of a perfect passive participle and a noun in the accusative is peculiarly Vergilian. The accusative is 'retained' after the passive voice. In the active this example would be *tempora evinxerunt*, 'they bound their temples': in the passive it becomes *tempora evincti*, 'they were bound their temples', i.e. 'with their temples bound', or 'having their temples bound'. This construction is used in imitation of the Greek 'middle' voice which expresses what a person does to or for himself: cf. XI, 35 *crinem solutae*, 'having their hair loosened'.

287-8 There are two groups of *Salii*: the young men dance, accompanying the singing in all probability with mime and gesture; the old men sing the hymn to Hercules in which the wonderful stories of his childhood, and his Labours (*laudes*) which he accomplished when a man (287-304), are told.*

287 **hic...chorus, ille** (*chorus*) - In apposition to *Salii* (285).

288 **Herculeas** - Adjective from the name, meaning 'of' or 'belonging to', agreeing with *laudes* in the preceding line. The use of adjectives from proper nouns is common in Latin and especially so in poetry.

ferunt - 'tell (of)'.

288-9 Translate in this order: *ut prima monstra novercae manu geminosque angues premens eliserit*.

288 **prima** - 'the first which he slew were...'.

novercae - The monsters were sent by his stepmother: the genitive is possessive.*

289 **geminosque...angues** - the explanatory use of *-que*: the *monstra* are the *angues*: translate as 'even'.

eliserit - Perfect subjunctive in an indirect question introduced by *ut* (288) in primary sequence after *ferunt* (288): cf. *disiecerit* (290) and *pertulerit* (293).

290 **idem** - Used idiomatically to make an added statement about a person already mentioned, in this instance about Hercules, 'how he *also*...'.

NOTES (VIII, 291–4)

291 **Troiamque Oechaliamque** - in apposition to *urbes* in the line before: both were sacked by Hercules.*

Troiam - Hercules sacked Troy when King Laomedon refused a reward he had promised him for killing a sea-monster which had been ravaging the countryside.

Oechaliam - Hercules sacked this city because the king, Eurytus, refused to give him his daughter Iole in marriage, although he had already promised that she would be his bride.

mille - A poetic usage meaning 'many' just as 'thousand' is often used in English.

labores - The myths about Hercules usually tell of twelve Labours, but in the hymn addressed to him which follows (293–304) only some of these are mentioned, together with others not contained in the twelve.*

292 **sub** - Put before *rege* in translation: 'in the power of'.

fatis Iunonis - 'by Juno's will'. Juno hated Hercules because he was the son of her husband, Jupiter, and a mortal, Alcmena. Influenced by consuming jealousy, Juno caused Hercules to become the slave of the king Eurystheus, who treated him harshly and set him twelve seemingly impossible tasks to do. Hercules, however, carried them all out successfully.*

293 **tu** - Notice the abrupt change to the second person which brings with it a feeling of reverence and awe. Some of Hercules' great deeds have already been related (287–93), but now it seems as if the hero is present with his worshippers.

nubigenas...bimembres - The first Labour, but not one of the usual twelve, mentioned in the hymn.* Compound adjectives such as these, made up of two, or sometimes more, words, are used by Vergil quite frequently, to give added effect. These emphasize the strange birth and form of the centaurs Hylaeus and Pholus, named in the next line. Centaurs were fabulous creatures, with the body of a horse and the head and shoulders of a man: they were the children of Ixion and Nubes (whose name means 'a cloud').

invicte - Vocative, addressed to Hercules. This was the cult title given to him in the ritual observed at the *ara maxima*.*

294 **manu** - Cf. 289, where the word is used to emphasize his

NOTES (VIII, 294–7)

strength as a child. The repetition in this line **again** stresses his physical powers.

Cresia - With *prodigia* in the next line: poetic plural. The 'Cretan monster' was a fine bull belonging to Minos, king of Crete. The bull became mad, but Hercules saved Crete from its ravages by carrying it off on his shoulders and bringing it back alive to Eurystheus.

mactas does not strictly refer to the Cretan bull (*Cresia... prodigia*), because Hercules did not kill it; but he did kill the Nemean lion which is mentioned in the next line: the two monsters are grouped together as being victims of Hercules' strength. The present tense is not so much a historic present as intended to draw attention to the action of the verb rather than the time when it was performed. Hercules slew the monsters and remains for all time as 'the slayer'. The verb almost bestows on him his title.

295 Translate in this order: *vastum leonem sub rupe Nemeae (mactas)*. This is the first of the conventional 'Twelve Labours'. Nemea was a valley near Thebes in eastern Greece. Hercules was sent by Eurystheus to slay a lion which was causing havoc in the countryside. When he found that it could be killed neither with bow and arrows nor with his club, he strangled it. Hercules is always shown in statuary wearing a lion-skin, but this is not the spoils of the Nemean lion, but of another slain for another Grecian king.

296 **Stygii...lacus** - 'The Stygian Lakes' were the rivers of Hades, the lower world of the dead (i.e. 'hell'). One of them was called *Styx*. The adjectival form of the name is used for a general description of them all.

ianitor Orci - The three-headed dog Cerberus, who was the watch-dog at the entrance to Hades. Eurystheus sent Hercules to fetch him. He was allowed by Pluto, king of the underworld, to drag Cerberus away, show him to Eurystheus, and then to take him back. This proved to be the most difficult of all the Labours. This is the usual 'Twelfth Labour'. With *ianitor* as the subject, supply *tremuit* from *tremuere*. One verb often does the work of two in Latin.

297 **super** comes in meaning in front of *ossa*. In poetry prepositions are frequently placed after the word with which they are associated (cf. 292).

semesa agrees with *ossa*.

NOTES (VIII, 297–302)

antro - The den in which Cerberus lived like a watch-dog at the entrance to Hades. Notice the harsh sounds in this line suggestive of a dog's bark, especially *-ans, antro*.

298 **facies** - 'shapes [of terror]'; supply *terruerunt* from the same verb in the line, and cf. note on 296.

Typhoeus - A fire-breathing giant with a hundred heads who fought against Jupiter and all the gods. When he was conquered he was buried beneath the volcano Aetna, in Sicily. In the traditional myths there is no mention of a fight between this giant and Hercules, but he is the type of monster with whom Hercules is associated in his various Labours, and so it is natural that Vergil should mention him.

299 **arduus arma tenens** - 'brandishing his arms on high'. *arduus* is used both to describe the giant's great height, and in an adverbial sense to give an impression of his threatening appearance.

rationis - Either the genitive or the ablative case is used with the verb *egeo*. The genitive expresses the sphere within which the lack is felt.

egentem - Present participle, agreeing with *te*.

299–300 'nor did the snake of Lerna coil about you with its shock of (snakes') heads, (and find you) lacking in resourcefulness'.

300 **Lernaeus...anguis** - A water-snake, often called the 'Hydra', with nine heads, which had its dwelling at Lerna near Argos in south-eastern Greece. If one of the heads was cut off, two others grew in its place. This happened until Hercules thought of burning them off and in this way slew the monster. He is described in 299 as not lacking resourcefulness because he thought out this plan of action. This is the conventional Second Labour.

turba - Ablative of description.

301 **Iovis** - Genitive of *Iuppiter*. Hercules was the son of Jupiter and a mortal woman, Alcmena (note on 292).

proles - Vocative, addressed to Hercules.

decus addite divis - 'added (as) a glory to the gods'. Although Hercules was a 'hero' rather than a god, he was sometimes regarded as semi-divine and was worshipped as a god.

addite - Perfect participle of *addo* in the vocative.

302 'both us and your sacrificial rites (i.e. done in your honour)

approach with propitious step'. Hercules is called on to be present with his worshippers.

dexter - Adjective, 'on the right side', agreeing with (*tu*) the subject of *adi*: the Romans believed that events which took place on the right side were lucky and that, conversely, those on the left were unlucky.

adi - Imperative from *adeo*.

303 **Caci** - They add (*adiciunt* in the next line) the tale of Cacus because it is a new 'Labour' and belongs peculiarly to Evander and his people. The rites which are being celebrated at the *ara maxima* have been established especially to honour it. All the rest which have been told are old Greek stories, but Cacus brings us to the heart of Rome. For the story see the caption in the text for lines 184–279.

super omnia implies that the slaying of Cacus is marked with special honour. It is thought that in origin Cacus was a primitive Italian fire-god worshipped from the times of the early settlement on the Palatine. On the side of the hill near the Tiber an old flight of steps provided a steep but short cut to the top and was called the *scalae Caci*. It must be noticed, however, that Vergil places Cacus' cave on the Aventine.

305 **nemus** - Where the sacrifice is in progress (108).

colles - The three most important hills of Rome, which stood on three sides of the Forum Boarium, the Capitol, the Palatine and the Aventine.

306 **urbem** - Pallanteum.

306–7 **divinis rebus...perfectis** - Ablative absolute.

307 **ibat** - The imperfect tense in this verb and in those in the following lines, *tenebat* (308), *levabat* (309), gives an effect of continuous action. We seem to watch, as in a moving picture, Evander, Aeneas, and Pallas walking along together. This is the true use of the imperfect in narrative.

obsitus - Literally 'sown with', hence 'covered with' or 'full of'. Evander moves slowly because of his age.

308 **natum** - Pallas.

tenebat - 'held [them] close', because he had many things to tell them.

309 **ingrediens** agrees with *rex* (307).

310–12 As they walk along and then climb up a path to the top of the Palatine Hill, they look at the view which extends over all the site of future Rome. Now they see only the wooded

NOTES (VIII, 310–16)

hills, grassy slopes and valleys which were there before Rome was built, but there are some remains left by the primitive men who lived there before Evander came.*

310 **faciles...oculos** - 'quick glances'. Aeneas gazes eagerly here and there over the scene.

311 **capitur** - 'is fascinated'. Aeneas, as founder of the Roman race, might well be fascinated with the sight of the hills and valleys where Rome of the future was to arise.

singula - Distributive adjective agreeing with *monimenta* (312).

laetus - Adjective used adverbially.

311–12 'one by one he joyfully questions and hears about the memorials of earlier men'. Primitive men had lived here before Evander came, and left traces of their old settlements. Evander tells the tale as they go along.

312 **virum** = *virorum*.

313 **conditor** - Evander is given this title because he was the first to build a settlement on the Palatine Hill. Vergil does not mean by this that he was the founder of Rome, because Romulus, who came later, was regarded as the real founder (48 and note, above).

314–36 - Evander recounts the primitive history of Latium and the site of Rome. There had been three ages of man from the beginning until Evander's own time: (1) the age of the 'aborigines', men born from trees, who lived the most primitive kind of life, knowing nothing of agriculture and having no settled homes; (2) 'the Golden Age' when the god Saturn ruled over mankind. He introduced law and peace and gave the name 'Latium' to the land; (3) a less glorious age spoilt by war and greed when invaders entered the land. The last of these was Evander.*

314 **indigenae** - From *indu*, the old form of *in*, and root of *gigno*; 'native' or 'aboriginal'.

315 It was a common belief among the Greeks and Romans that the earliest men on earth came out of trees and stones.

trunics...robore - Ablatives of source or origin.

duro suggests that they were especially hardy since they were the offspring of strong oak-trees.

316 **quis** - Dative plural (possessive) of *qui*.

mos - Roman law in certain aspects was founded on *mos*, that is, a traditional way of life approved and found good, handed

NOTES (VIII, 316–20)

down from one generation to another. It was a necessary condition for social life.

iungere tauros - The Romans used oxen for ploughing and as draught animals, just as the Italians still do today.

316-17 'who had neither custom nor an ordered way (of life) and did not know how to yoke oxen nor store their wealth, nor make thrifty use of their gain': these primitive men had not discovered how to cultivate the land. The use of the prolative infinitives *iungere, componere, parcere* with *norant* (317) is poetical.

317 They did not know how to store up their food, but lived from one day to another on whatever they could find.

norant - Shortened pluperfect (*noverant*) from *nosco*. The verb means 'get to know' or 'learn', hence the perfect means literally 'have got to know', 'have learnt', i.e. 'know'; the pluperfect carries the meaning therefore of 'knew'.

parto - Perfect participle of *pario* used as a neuter noun, in the dative with *parcere*.

318 'but branches and hunting which provided rough food (lit. "rough with food") sustained them'. They lived on wild fruit and berries, and the raw flesh of hunted animals.

victu - Ablative of the fourth declension noun, *victus*; connected with *vivo* (supine *victum*), 'live'.

319 **primus...venit** - 'was the first to come'. This is a special use of *primus* when it agrees with the subject.

Saturnus - Saturn, the god of sowing; the name is connected with *sero*, 'sow' (supine *satum*). He was thought to have taught agriculture to the first inhabitants of Italy, and to have been the father of Jupiter.

Olympo - A mountain in Thessaly in northern Greece where the gods were thought to have their home.

320 The Romans identified *Saturnus* with the Greek god *Cronos* who was expelled from his kingdom by his son Zeus (*Iuppiter* in Latin). The story, as far as Vergil tells it here, is that Saturnus (or Cronos) then fled to Italy and, as a result of the agricultural skills which he introduced into the country, brought in the 'Golden Age' (324-5).

Iovis - Genitive of *Iuppiter*. According to the myth, Saturnus devoured his sons as soon as they were born of Ops his wife, because one of them was destined to oust him from his kingdom. Jupiter was saved by Ops, who substituted for the

NOTES (VIII, 320–6)

child a stone which Saturnus devoured instead. So it was that Jupiter grew up eventually to drive out his father Saturnus from his kingdom.

321 **is** - Saturnus.

indocile usually carries the meaning 'unteachable', but here is merely passive, 'untaught'.

montibus altis - Local ablative without a preposition: natural places for primitive man in which to take cover and refuge, and to hunt in the forests.

322 **composuit**, 'he settled', implies that he both taught them (cf. *indocile* in 321) and gathered them into communities (cf. *dispersum* in 321).

322–3 An imagined derivation of the name Latium, after Vergil's manner. He often attributes the origin of a place-name to some myth or some characteristic associated with it. Here the explanation is that Saturnus gave the name to the country because he had found safety there: hence 'Latium' means the 'Land of Hiding' and is derived in Vergil's fancy from *lateo* ('lie hid'). Latium is the country which stretches from the Tiber south-eastwards between the Mediterranean and the Apennines and includes Campania. Within its extent is the plain on which Rome stands. It is the setting of the last six books of the *Aeneid*.

323 **maluit** - He preferred to call the country Latium rather than to call it after himself: in a later age it received his name (329).

latuisset - From *lateo*. The subjunctive is in indirect speech (*oratio obliqua*) depending on *maluit*. The sense is 'Saturnus named it Latium, because (as he said)...'. This was the reason he gave when he named it.

324–5 Translate in this order: *saecula quae aurea perhibent sub illo rege fuere*. The reign of Saturnus brought in the 'Golden Age'.

324 **perhibent** - A general use of the third person plural; to be compared with the English idiom, 'they say....'.

325 The alliteration of the letter *p* with its soft, gentle sound imitates the quietness of peace.

326 By contrast with the preceding line, the alliteration in the *d* sound gives an impression of harshness, suggesting the disquiet of war and greed, perhaps even carries a hint of reproach in the very sound.

deterior agrees with *aetas*.

68

NOTES (VIII, 326–32)

decolor - 'faded' or 'dim'. Vergil perhaps intends to represent the colour of a baser metal such as bronze or iron which would seem less bright when compared with gold. From the literal meaning of 'off colour', 'tarnished', the word comes to mean 'degenerate' or 'depraved'.

327 **habendi** - The use of the gerund as a verbal noun is common in Vergil. This example in the genitive is associated with *amor*.

328 **manus Ausonia...Sicanae** - Names of tribes which the Romans believed inhabited Latium in primitive times. The *Ausones*, from whose name the adjective *Ausonius* is formed, were thought to have been among the earliest invaders of Latium. They seem originally to have spread over all central Italy, but later to have settled in part of the west coast of Latium and all Campania, to the south. The Sicani were another early invading tribe. It was believed that they first came from Spain to Sicily and from there to Latium.

venere - Perfect tense (*venerunt*).

329 'and often the land of Saturn laid down its name'. The name *Latium* given by Saturn began to go out of use, since the 'Golden Age' was passed. *Saturnia* is an adjective from the name *Saturnus* (note on 288).

330 **immani corpore** - Ablative of description.

Thybris - Several stories were centred around the river in explanation of the origin of its name. It was said that *Thybris* was a mythical king of the Etruscans who fell fighting near the river and gave it his name; in another legend he was said to be a robber king who had his haunts in the neighbourhood and used to waylay people as they ferried across; a third told how an Alban king, *Tiberinus*, was drowned in the river and how it was called *Tiberis* after him. It is probable that the name *Thybris* is the Etruscan equivalent of the Latin *Tiberis*. The Etruscans lived in the country on the right bank of the Tiber and exercised a strong control over the river in early times.

331–2 'from whom (i.e. "after whom") afterwards (we) Italians called (our) river *Thybris* by his name'.

post - Used here as an adverb: *postea* is more usual in this meaning.

Itali - The use of this name on the lips of Evander shows that Vergil is looking forward into history and connecting him and

NOTES (VIII, 331–6)

his people, the Arcadians, who were Greeks, in a general way with the Italians of the future.

332 **Albula** - A name for the Tiber apparently older than *Thybris* or *Tiberis*. It was thought that the name was changed from *Albula* after the drowning of Tiberinus. It is probably connected with an old root *alb* for mountain, hence 'the mountain river' (note on 330).

333–6 The main structure of this sentence is: *me pulsum... sequentem Fortuna...et...fatum...posuere..., matrisque... monita et deus...(me) egere*. *Fortuna* and *fatum* are both subjects of *posuere*.

333 **pulsum** - Legend told how Evander slew his father at the bidding of his mother. Then, being guilty of bloodshed, he fled from Greece and reached Italy. There he built a settlement on the Palatine Hill and called it Pallanteum, long before the founding of Rome.

pelagique extrema - 'the farthest ends of ocean'. Evander had sailed westward from Greece over a sea of which the limits were unknown to him. The voyage would, in his experience, seem like going to the end of the world. The neuter plural of an adjective (*extrema*) with the genitive (*pelagi*) illustrates a fairly frequent and peculiarly Vergilian usage in the description of places. The effect is to stress the meaning of the adjective.

334 For sound alone one of Vergil's finest lines. The grandeur of the adjective *omnipotens* and the long drawn out sound of the six-syllable word *ineluctabile* fit the unyielding relentlessness of the powers which rule Evander's destiny. Driven on by them, even though he was not primarily responsible for his murderous deed, he could not escape the necessity of making atonement. In the very sound of *ineluctabile* there is a sense of yearning against implacable fate.

335 **posuere, egere** - Perfect tenses.
tremenda - With *monita* in the next line.

336 **Carmentis Nymphae** - Genitive in apposition to *matris* in the preceding line. Carmentis was a native Italian goddess of prophecy whom Vergil chose to identify with the mother of Evander and thus to weave her name into Evander's story (337–41). She was called in the Greek legend Nicostrata.*
auctor Apollo - The god of prophecy gave Carmentis her prophetic powers (*monita*). He was her 'authority'.

NOTES (VIII, 337-44)

337-9 Translate in this order: *monstrat et aram et portam quam Romani Carmentalem nomine memorant*.

338 **Carmentalem**, although in the singular, agrees with both *aram* and *portam*.

339 **priscum** - *Priscus* always carries the implication of something belonging to the earliest times: it has nearly the same meaning as our 'primitive' when we mean 'early'. Vergil shows by using this adjective that he considered the altar and gate to be older than Rome.

honorem - Either in apposition to *portam* or more probably an accusative in apposition to the preceding sentence. The actual naming was done in her honour.

340 **cecinit** - From *cano*.

quae comes in meaning before *cecinit*.

prima - 'was the first to...'.

342-3 After Evander and Aeneas have passed through the gate, they have in view the two most important of the seven hills of Rome. On their left is the *Mons Capitolinus* (*Capitolium*), often called the Capitoline or Capitol in English, and on their right the *Mons Palatinus*, the Palatine Hill, on which Evander has his home. Evander points to an eminence on the summit of the Capitoline and then to a cave in the side of the Palatine. Between the two is the valley which later became the Roman Forum.*

342 **lucum** - Vergil thinks of the Capitol as covered with forest growth in these early times before the coming of Romulus.

asylum - From a Greek word for a 'refuge'. The legend told how Romulus made a place of sanctuary on the Capitol where he received refugees and outlaws from the surrounding country. These people became the first inhabitants of the Rome which he founded.

343 **rettulit** - Lit. 'brought back', i.e. 'rendered', 'made'.

Lupercal - A cave at the foot of the Palatine in which it was said that a she-wolf nursed Romulus and Remus. The name is connected with *lupus* ('wolf'). It was regarded with great reverence in Augustan times and decorated with sculpture.*

344 '(the Lupercal) called after Parrhasian (i.e. Arcadian) custom (the place) of Lycaean Pan'. *Lycaeus* in the Greek is connected with the word for 'a wolf'. Vergil draws a comparison between a place in Evander's native Arcadia and the site of Rome.

NOTES (VIII, 344–52)

Panos - Greek genitive; *Lycaei* agrees with it in the Latin form. The Romans took pleasure in finding Greek equivalents for their own gods.

345 **nec non** - A device often used to mark an important statement.

Argileti - A quarter in Rome through which a street ran from the Quirinal Hill to the Forum. We are to think of it in Evander's time as a wooded dell through which a stream runs down into the valley of the Forum.*

346 **testatur** - Evander calls the place to witness that he was innocent of the death of Argus. Although *Argiletum* probably means only 'a place where clay is dug', Vergil takes it to mean 'the death of Argus' by suggesting a fanciful derivation as in the case of Latium (322–3 and note, above). Argus had once been the guest of Evander, but when he plotted against him Evander's loyal subjects put him to death without the king's knowledge. No wonder that looking at his place of burial Evander declares his own innocence.

347 Next they climb up the Capitoline Hill, and see the Tarpeian Rock, which was a place of execution, named after the girl Tarpeia who betrayed the hill to the Sabines.*

348 **aurea nunc** - A reference to the gilded roof of the temple of Jupiter Capitolinus which stood on the summit in Vergil's day.

349 **religio** - Not 'religion' in the modern sense, but 'fear', 'reverence', and such extended meanings as 'worship', 'observance of ritual'; 'holiness' or 'sanctity' in pagan cults. The word is connected with *ligo*, 'bind', and has to do with the 'binding' or compelling demands of fear and superstition.

350 **dira** - With *religio* in the preceding line.

tum - In the time of Evander.

silvam saxumque - The forest-covered rocky hill.

351–2 '"This grove," Evander said, "this hill with leafy summit, uncertain it is which god, (yet) a god inhabits."' Evander means to say that a spirit seems to haunt the hill-top. He is speaking about a time far anterior to the building of the great temple of Jupiter Capitolinus, but he goes on to tell how Jupiter has been seen in the grove by some of his own people (352–3). This is the spot where later the temple was built. Jupiter was thought therefore to have chosen the situation for his temple long before it was built.*

NOTES (VIII, 353–60)

353 **saepe** represents a condensed expression such as, 'as he often does'.

nigrantem agrees with *aegida* in the next line.

354 **aegida** - Greek accusative. This was an attribute of Jupiter, associated, because of its shape, with storm-clouds. He was thought to cause storms and to bring rain by shaking it: for this reason it is described as *nigrantem*.

355 **haec** - Deictic use of the demonstrative adjective, implying that Evander points to the ruins as he speaks.

duo...oppida - The mention of these two towns leaves no doubt that Evander and Aeneas have climbed to the top of the Capitoline. They look at old walls left by earlier inhabitants on this hill and on the Janiculum Hill which is across the Tiber, but cannot be seen until the spectator reaches the top of the Capitoline. It is quite possible that in Vergil's time there were some remains of old walls which were thought to be older than Rome.*

357 **hanc...hanc** - Deictic. The town on the Janiculum was founded by Janus, and that on the Capitoline by Saturnus. They were both gods. There was in Roman times a temple dedicated to Saturnus at the foot of the Capitoline, facing the Roman Forum. The hill itself was sometimes called after it *Saturnius*. The legend of an old town of the same name arose from this.

358 **fuerat** - The pluperfect tense stresses the great antiquity of the towns.

359–60 They walk through the valley, after descending the Capitoline Hill, where later the Roman Forum was to be, along a track which afterwards was the paved road called the *Sacra Via*. They then climb up the gentle incline on to the northern edge of the Palatine Hill to reach Evander's home.

ad tecta...Euandri - Because his poverty is stressed (*pauperis*) it is clear that Evander's house was no more than a group of huts made of wattle and reeds or straw. This kind of primitive dwelling was familiar to all Romans from the *casa Romuli*, the so-called 'house of Romulus', a sacred hut which was preserved on the Palatine Hill. Even at the present day the shepherds of the Roman Campagna (the country district outside Rome) live in reed huts of this simple type.

NOTES (VIII, 360–5)

360–1 From the higher ground of the Palatine Hill the view extends over the heart of Rome, for Evander and Aeneas then only grass-covered slopes and hollows. They look down into the valley of the Forum below and see the herds grazing peacefully. They are freed from fear now since the cattle-raider Cacus has been slain by Hercules. The ridge of the Oppian Hill (*mons Oppius*) and the southern edge of the Esquiline (*mons Esquilinus*) which rises beyond it are both grass-covered slopes, grazing ground for the cattle. In Vergil's time this was a fashionable residential quarter called *Carinae*, hence the adjective *lautae*, 'elegant'.

361 foro, Carinis - Local ablatives without a preposition.

mugire - The sound of the cattle lowing as they graze increases the feeling of peace and security and gives a vivid touch to the scene.

362 ventum - Supply *est*; this is an example of an 'impersonal passive', a construction in which the action of the verb is described without mention of the subject. It is only found with intransitive verbs, usually those of motion.

362–3 victor Alcides - Hercules had visited Evander's house after slaying Cacus and recovering the stolen cattle. For the patronymic *Alcides* see note on 103.

363 regia - Evander calls his poor home a king's house because it is his, but there is no kingly splendour about it.

cepit - 'received', 'welcomed'.

364–5 'have the courage, my guest, to despise wealth, and think yourself, too, worthy of (the hospitality once accepted by) a god, and enter (my house) not scornful of simple living'. These lines are among the most often quoted, and best loved, in the *Aeneid*. In them the Romans' admiration for hardihood, humility and the simple life is expressed. They were pleased to think of their early kings and heroes living the hard life, free of luxury, of a rustic community. In Vergil's own day the emperor Augustus had a palace on the very spot where Vergil puts Evander's home. Augustus, true to Roman tradition, had what was, compared with his position, a building quite without the magnificence and outward display of wealth which might have been expected. Evander's words to Aeneas contain an inner reference to Vergil's own day and to Augustus himself, who did not indulge in extravagance and in the riches on which he could have drawn.

NOTES (VIII, 364–456)

Aeneas, the founder, through his son Iulus, of the Augustan line, is called to despise wealth and is offered the hospitality of a humble house on the very spot where afterwards his great descendant was to have his unpretentious house.

365 **deo** - Ablative with *dignum* in the preceding line. Besides referring to Hercules, about which there can be no doubt (363, *Alcides*), this word also indirectly refers to Augustus who was reverenced as a god even in his own lifetime.

veni - Imperative, as scansion shows (*vĕni*).

366 **fastigia** - 'pointed roofs' of the huts of Evander's home.

367 **ingentem** - In Vergil this word often carries a wider meaning than merely 'large', but, as well, 'remarkable', 'great' with some deep significance. Aeneas seems perhaps physically 'great' compared with the house which he enters, but he is 'great' because of his destiny, for as the ancestor of Augustus and the founder of the Roman race he enters a king's house on the site of future Rome. He is 'remarkable' for all he represents to a Roman, and at the same time 'great' enough in another sense not to despise the poor huts of Evander's settlement. He is the man of destiny and of power.

367–8 'placed (him) (i.e. made him lie down) on strewn leaves and on the hide of a Libyan she-bear'. Aeneas has for a bed a heap of leaves covered with a bear-skin. This is a countrified kind of bed in keeping with Evander's poverty. The covering is made more picturesque with the name 'Libyan'.

368 **effultum** - Perfect participle from *effulcio*: agrees with *Aenean* in the preceding line.

foliis, pelle - Instrumental ablatives.

454 **pater...Lemnius** - Vulcan, the god of fire, who is forging arms for Aeneas with all haste at the command of Venus (see caption for lines 370–453 and 608–731 in the text). This god was called after the island where he fell to earth after being thrown out of heaven.

455 **suscitat** - 'wakened him (and called him out...)'.

alma - 'fostering', that which gives life, from *alo*, 'nourish'.

456 **matutini...cantus** - Subject of *suscitat*, to be supplied from *suscitat* in the preceding line. It is not uncommon for one verb to do the work of two in a compound sentence, even where one is singular and the other plural.

volucrum - Evander wakes early at the first call of the birds,

NOTES (VIII, 456–64)

whether these are swallows twittering under the eaves, cocks crowing in their sheds, or the dawn chorus in general.

457 **tunicaque inducitur artus** - Lit. 'he is clothed as to his limbs with a tunic'. The passive is used here in the same way as the Greek middle voice, to describe an action done by the subject to himself.

artus - Accusative plural 'retained' after the passive voice: a common construction in Vergil with verbs meaning 'put on' of clothes or arms and armour.

458 **plantis** - Dative with *circumdat*.

459 **Tegeaeum** - Adjective from a place-name, 'from Tegea', a town in Arcadia. This word serves to remind us of Evander's Arcadian origin.

459–60 Evander wears a panther-skin over his left shoulder: his sword is fastened to the sword-belt which is held in place by a strap over the right shoulder (*umeris*).

ab laeva - 'on his left (arm)'.

retorquens = *retorta gerens*.

461 **nec non** - Cf. 345 and note.

gemini custodes - In apposition to *canes* in the following line. They are Evander's companions and the watch-dogs of his house.

limine ab alto - These words imply that Evander comes out of his own part of the house (probably a primitive hut) and goes to one of the other huts where Aeneas has slept that night.

462 **gressum...erilem** - 'their master's steps'. The dogs go along with him. *erilis* is the adjective of 'erus'.

canes - The Romans were fond of dogs and kept them for hunting, for companions, and also as watch-dogs. A mosaic from the entrance to a house in Pompeii shows a barking dog on a chain, and the words *cave canem*, 'beware of the dog'.

463 **sedem et secreta** - The separate room, or more probably hut, which Evander allotted to Aeneas as his guest: 'the quiet guest-room'.

464 'his mind full of their talk (of yesterday) and of the help he promised, (he) the hero-king'.

heros - A tribute to Evander. On first recognizing Aeneas as a kinsman he promised his help (171); now, in keeping with the character of an upright king and one faithful to his word, he is eager to fulfil his promise. Evander perhaps also merits this title because of his former exploits in war (563 and notes, below).

465 **matutinus** agrees with Aeneas, 'rising early', a somewhat unusual use of an adjective denoting time in relation to a person, perhaps in imitation of a common Greek idiom in this connection.

agebat - The imperfect of narrative, used for vividness.

466 **huic...illi** - Evander and Aeneas: possessive datives. Evander is accompanied by his son Pallas, and Aeneas by his faithful companion Achates.

ibat - 'went (to meet each other)'.

468 **licito...sermone** - The deponent *fruor* regularly governs the ablative case. Evander and Aeneas are at last (*tandem*) able to talk privately together: on the day before they had been among the crowd of worshippers until nightfall.

469 See note on 41.

haec - Neuter plural, 'these words'; supply *inquit*.

470 **maxime** - Vocative, agreeing with *ductor*; addressed to Aeneas.

quo sospite - The antecedent to *quo* is *ductor*: the phrase is an ablative absolute.

470-1 'since you survive, I for my part shall never admit that the wealth and realms of Troy are vanquished'. Evander declares that in the person of Aeneas Troy lives on. The further implication is that Troy will rise again on Italian soil.*

471 **equidem** - Adverb, often used to emphasize *ego*.

472 **nobis** - Possessive dative.

ad belli auxilium - 'for help in war'; *belli* is a defining genitive, explaining the sphere in which help is needed.

pro nomine tanto - 'in view of our great name'. Evander explains that in spite of the fame of the Arcadians and his own distinction, his resources are small. All through the book his poverty is stressed: cf. especially 364-8, 455.

473 **Tusco** - The Tiber was often called 'Etruscan' or 'Tuscan' because its course formed the boundary between Roman and Etruscan until the Roman conquest of Etruria.

amni - Old form of the ablative.

474 **Rutulus premit** - The Rutulians (*Rutuli*) under the leader-

ship of their prince, Turnus, are already at war with Evander because of the tyrant Mezentius who has been driven out of Agylla (478–95). Evander favours the rebels, whom he considers to have a righteous cause. Turnus has taken up the cause of Mezentius and at the same time has been at war with the Trojans since he attacked their camp at the mouth of the Tiber during Aeneas' visit to Evander (*Aen.* IX, 1 ff.). The territory of the Rutuli was the district around their tribal centre called Ardea, on the plain of Latium about twenty miles to the south-east of Rome. A village of that name still exists on the site to this day.

475 **ingentes populos** - The Etruscans, who were a very powerful people in early times: *ingentes* suggests might rather than size (note on 367 above).

regnis - Twelve city-states made up the Etruscan league: Evander implies all the forces which can be mustered.

476 **quam...salutem** - In apposition to the preceding lines. What Evander proposes will bring *salus*.

fors inopina - The unexpected chance is that the Etruscans are looking for a leader.

477 'You have come (and are) here at the call of destiny.' The present tense stresses the actuality of Aeneas' presence: he has been brought by the inexorable demands of fate. All through the *Aeneid* we find it stressed again and again that Aeneas was led by fate, by a fixed destiny stronger even than the will of the gods, to reach Italy and to be victorious in war so that he might become the founder of the Roman race. In this is to be seen a reflection of the Roman belief that the greatness of the nation and her dominions and conquest of the world were the outcome of the will of a destiny which could not be turned aside.

te...adfers - Lit. 'you bring yourself here', i.e. 'you have come'.

478–80 'Not far from here, founded on an ancient rock, there is inhabited the site of the city of Agylla, where once (i.e. long ago) a Lydian race distinguished in war settled on the ridges [and called them] Etruscan.' The Etruscan city called in the Greek Agylla, and in the Etruscan Caere, was about twenty miles to the north-west of Rome and about five miles inland from the shore.*

481 **florentem** - 'after flourishing through many years': the

NOTES (VIII, 481-95)

present participle emphasizes the continued state of prosperity which it was, and had been, enjoying when Mezentius became the tyrant.

deinde refers back to *multos...annos*.

483 **memorem** - Deliberative subjunctive, used when a person wonders what to do: in this passage it is used for rhetorical effect. By suggesting that perhaps he will pass over the very subject of which he wishes to speak, Evander draws attention to it: his intention is to stress Mezentius' cruelties, not to leave them untold.

484 **generi** - Dative of *genus*.

reservent - Subjunctive of a wish or prayer: Evander prays that Mezentius and his descendants may suffer for the cruelty he has inflicted on others.

487 **genus** - Accusative in apposition to the preceding lines.

488 **longa** - 'long drawn out', 'lingering'.

489 **infanda furentem** - The present participle agrees with 'him' (i.e. Mezentius) understood. *infanda* is the neuter plural of the adjective used as an internal accusative in an adverbial sense: the literal meaning is 'raging horrible ragings', which comes to mean 'raging horribly'. Translate, 'the madman committing horrible atrocities'.

490 **-que...-que** = *et...et*.

ipsum - Mezentius.

491 **socios** - 'his confederates', those who supported Mezentius and connived at his cruelties.

fastigia - The roof of Mezentius' house: probably in those early days of thatch so easily ignited.

492 **Rutulorum** - With *agros*.

493 **confugere, defendier** - Historic infinitives; these are regularly used to describe vividly a succession of actions and are equal in meaning to the imperfect tense in narrative.

defendier - Old form of the present infinitive passive. Vergil uses archaisms to give an antique flavour to the poem.

hospitis - In apposition to Turnus who gave refuge to Mezentius and so became his host. The word can mean either 'host' or 'guest' according to the context (cf. 364).

495 **regem** - Mezentius. The Etruscans demand that he should pay the penalty for his crimes.

praesenti Marte - 'with immediate war'. They did not wait

79

NOTES (VIII, 495–503)

to try other ways, but at once went to war against Mezentius and Turnus.

496 Evander proposes to put Aeneas at the head of the Etruscans who have revolted against Mezentius. They need one who can lead them all, together with the Arcadians. Evander is influenced to do this because it is prophesied that they must wait for a foreign leader: he recognizes in Aeneas the promised one.

Aenea - Vocative.

497 **fremunt** - The subject is *puppes*. The feelings of those who man the fleet are transferred to the boats as if they too are impatient for the battle. Such a turn of expression is characteristic of Vergil's style (91–3 above).

puppes - 'ships'. A part of a ship is often used for the whole. The ships are those of the Etruscan fleet which is ready to sail from the Etruscan shore north-east of the mouth of the Tiber (*toto litore*). There were several Etruscan ports along this coastline. This fleet later sails down the coast to the mouth of the Tiber with Aeneas at the head.

498 **longaevus haruspex** - 'a prophet of great age'. We must think of him as an Etruscan priest of age and rank whose words command respect, although his name is not given. He speaks to the Etruscan troops who wait for battle but as yet have no leader.

499 **canens** - *Cano* is often used for the uttering of prophecies, because they were often chanted, especially when given by oracle.

Maeoniae - Old name for Lydia in Asia Minor, supposed to be the Etruscans' original home.

500-1 'Flower and glory of fighting men of old, (you) whom righteous indignation arouses against the foe and (whom) Mezentius fires with just anger.' The alliteration of *v* sounds in 500 suggests the anger they feel.

502 Supply *est* with *fas*. The Etruscan race, glorious, wealthy and splendid, proud in its ancestry, is too great to have an Italian (a mere native of the land) to be its leader. They should have a leader of distinction to match their greatness.

nulli - Dative, with *Italo*.

503 **duces** - Poetic plural: the Etruscans need one supreme leader.

NOTES (VIII, 503–12)

503–4 Take in this order: *tum Etrusca acies resedit hoc campo monitis divum exterrita*.

monitis - From the noun *monitum*.

505–6 Tarcho became ruler of Agylla after the flight of Mezentius: he does not feel equal to the task of leading the Etruscan cities which have risen as one against Mezentius, so has sent the *regalia* which belong to a king to Evander in the hope that he will be at their head.

507 succedam...capessam - Subjunctives expressing indirect command: supply 'begging me to...'.

508–9 'But old age, slowed by the chill (of infirmity) and tired out by the passing years, grudges me the power (of leadership), and my bodily strength too slow for brave deeds.' Evander does not feel strong enough, because of his great age, to go into battle. These lines summarize graphically the effect of age on a person, at the same time showing that sympathy which Vergil felt for the weak and suffering. His humanity is perhaps nowhere more clearly recognized than in these lines.

508 mihi - With *invidet* (509).

gelu - A metaphor describing the freezing effect of old age on the body: it numbs activity.

saeclis - Literally a 'generation', which is usually calculated to be thirty years: hence 'passing of time'.

510 natum - Pallas. Evander would have urged his son to put himself at the head of the Etruscan band, had not the seer pronounced that the leader was destined to be a foreigner. Pallas was half Italian by birth, although on Evander's side he was foreign.

matre - Ablative of origin. *Sabella* probably means 'Samnite'. Some scholars would call her 'Sabine'.

510–11 exhortarer, traheret - Imperfect subjunctives used to express what is the opposite of fact in present time—Evander implies 'I should now be urging him...did he not...' (but the fact is that he does...).

ni mixtus...traheret - 'if it were not that, mixed in descent from a Sabellian mother, he draws from this a portion of his fatherland'.

511–12 'You, to whose years and race alike the fates show favour.' Aeneas is younger compared with the old man Evander and older compared with Pallas. The Trojan race is under the special favour of destiny because it is the ancestor of the

NOTES (VIII, 511–23)

Roman race. Evander recognizes both these facts and so calls on Aeneas to be the leader: he is a foreigner and he is able to undertake the task, both by distinction and greatness.

512 Aeneas is demanded by the gods because of his great destiny to found the Roman race: in this idea we see a reflection of the greatness of Rome herself.

513 **ingredere** - Imperative. A very impressive word sometimes used of gods: the underlying idea is 'come forward in your might'.

Teucrum...Italum - Genitive plural.

fortissime - Vocative.

514 **hunc** - Pallas, who is at Evander's side. Evander turns to Pallas as he speaks and perhaps takes his arm or brings him forward.

spes - Plural.

solacia - Pallas is Evander's sole support because his wife is no longer alive and he is his only son.

nostri - Objective genitive of *nos*: plural pronoun often used for the singular. The genitive of the pronoun is used rather than the possessive pronoun which would be usual in prose, for effect and emphasis. Although Pallas is Evander's only hope and support, he is willing to let him go.

515 **Pallanta** - Greek accusative.

magistro - With *te*.

515–16 **tolerare, cernere** - Infinitives with *adsuescat*, a normal construction.

517 **adsuescat** - Subjunctive expressing command in the third person: cf. *miretur*.

518 **huic** - Pallas.

519 **lecta** - From *lego*.

Pallas - Supply *dabit* from *dabo*. The name is put at the end of the line for effect. In this Evander's love for his only son is reflected: there seems to be a hint of the tragedy to come in the lingering, lengthened sound at the end of the line. His thoughts seem to dwell on the very name.

520–3 The moment that Evander had finished speaking, a sign was given from heaven. *vix...erat* looks on to 523 and in reality goes with it in construction, but this is interrupted by *tenebant* and *putabant* so that the sentence becomes irregular. The straightforward statement is that Evander had scarcely finished speaking when a sign was given, but for vividness and deeper expression there is an interruption in the construction while their feelings are described: (*defixique...putabant*). The mention of the sign is then attracted into a construction with the verbs *tenebant*, *putabant*, which is an irregular conditional sentence. The apodosis contains two verbs in the imperfect indicative, *tenebant*, *putabant*, and the protasis introduced by *ni* contains a verb in the pluperfect subjunctive, *dedisset*, whereas in the regular form of conditional sentences the same tense and mood are used in both clauses. In relation to *dedisset* there is an underlying suppressed thought closely bound up in *tenebant* and *putabant* which would contain a pluperfect subjunctive if put into words: 'they held their heads downcast...and were pondering (and would have continued doing so)' had not Venus given the sign which put an end to their anxiety. This is an example of a 'suppressed apodosis'.

520 **defixi ora tenebant -** *defixi* goes in meaning equally with *ora* and *tenebant*: 'being downcast (in thought), they held their faces (downcast)'.

523 **Cytherea -** 'The lady of Cythera', i.e. Venus; one of the epithets often given to her, from Cythera, a Greek island, which was one of the centres of her worship.

caelo...aperto - 'in a clear sky': local ablative. The Romans believed that the gods showed their will by sending thunder and lightning, and that when this happened in a cloudless or bright sky the signs were especially favourable.

524 **improviso -** adverb. 'Lightning flashed in the heavens.' The Latin, as often, uses 'ab' for the starting point.

525 **ruere -** Infinitive with *visa* (*sunt*): not 'to fall' so much as 'to shake', describing the impression given by a loud crash of thunder.

526 **Tyrrhenus tubae...clangor** = 'the sound of the Tyrrhenian trumpet'. The epithet is transferred from the trumpet to the sound, after Vergil's fondness of inverting phrases to give added flavour to the meaning. The trumpet was especially connected with the Tyrrhenians (i.e. the Etruscans) and is here used for effect, but may carry an underlying thought of the help that is to come from Etruria.

527 **iterum...iterum -** The thunder is heard three times, constituting a complete omen in Roman superstitition.

528 **inter nubem -** The arms made by Vulcan for Aeneas at the

NOTES (VIII, 528–40)

command of his mother Venus are seen shining in the sky: they seem to float suspended on a cloud.

529 **pulsa tonare** - A verb of hearing is to be understood from *vident*. See the caption in the text for lines 370–453, 608–731.
pulsa - From *pello*.

531 **agnovit sonitum** - Aeneas recognized the sign as one sent by his mother, the goddess Venus. The vision of heavenly armour was a token that the armour which had been forged for him by Vulcan the god of fire was completed. War was certainly to be intensified, but her reassurance was for his comfort.

532 **ne...ne quaere** - Poetic use of *ne* with the imperative to express a negative command: the equivalent in prose of *noli* with the infinitive. The repetition of *ne* is for emphasis, to show that Aeneas' entreaty is urgent.
profecto repeats the meaning of *vero*, 'indeed'.

533 **ferant** - Subjunctive in an indirect question after *ne quaere* in the preceding line.
Olympo - 'from Olympus', the home of the gods.

534 **cecinit** - Cf. 499 and note on *canens*.
missuram - Supply *se...esse*: regular construction. Vergil does not tell us when Venus made this promise.
diva creatrix - Venus, his goddess mother.

535 **ingrueret** - Subjunctive in a conditional clause in *oratio obliqua*: in direct speech this would be the present subjunctive in a condition 'possible of fulfilment'.
Volcania - Epithet from the name *Volcanus*: translate 'made by Vulcan'.

536 **laturam** - From *fero*: for the construction cf. *missuram* (534) and note.
auxilio - Predicative dative. This is more usually associated with the verb 'to be', and expresses 'the end in view': in this example translate 'to help us'.

537 **Laurentibus** - A name for the inhabitants of the coastal district of Latium, over whom Latinus rules, but used here in a wider sense for the native peoples of Latium with whom Aeneas and the Trojans are at war and with whom he now knows the struggle will be severe and inescapable.

537–40 Aeneas' sorrowful outburst at the thought of war seems to suggest that he pities even his enemy. The use of the second person when he speaks of Turnus emphasizes his grief.

84

NOTES (VIII, 538–44)

538 **Turne** – Notice the dramatic effect of the sudden change to the second person.

539 **volves** – The subject is *Thybri* in the next line.

540 **Thybri** – Vocative.

poscant, rumpant – jussive subjunctives, expressing command: 'let them...'. There is an abrupt change, after the address to Turnus, back to the third person.

poscant acies implies, 'let them demand war, (if that is what they really wish)'.

foedera – An alliance had never been completed between the Latins and Trojans, but when Aeneas' messengers, sent ahead to reconnoitre the land, first made the acquaintance of Latinus he had welcomed them hospitably and shown his willingness to receive them on friendly terms (*Aen.* VII, 148–285).

541 **tollit** – The subject is *Aeneas*.

542 **Herculeis...aras** – An inversion after Vergil's manner. The altars did not sleep in the fires, but, *vice versa*, fires slept (i.e. had died down) on the altars.

542–3 'And first brings to life the altars slumbering with the fires of Hercules.' This is best explained by the conjecture that fire had been taken from the sacrifice at the *ara maxima* of the day before, and left to smoulder during the night on Evander's household altars. This was a familiar practice in ancient ritual, and since Hercules had been a guest in the house (*Aen.* VIII, 185–275), we may perhaps conclude that there was a shrine sacred to him within the house. Fire was a precious thing in the ancient world and replenishment from a sacrificial fire is an action which need not surprise us.

543 **hesternumque larem** – 'yesterday's hearth (fire)'. The *lar* was one of the Roman household gods, but the word can be used for the 'hearth', over which the *lar* presided. This is probably here the fire which Aeneas is carrying from Troy and which became the fire of Vesta.

penates – The two gods of the Roman household especially connected with the stores.

parvos suggests that they were the gods of Evander's small home.

544 **adit** – 'approached', to worship reverently.

bidentes – Two-year-old sheep. These were often offered in sacrifice. At this age the two front teeth are prominent.

NOTES (VIII, 545–53)

545 **Euandrus** - Rare form of the name *Euander*, used in this passage to avoid the repetition of sound in *pariter*.

An equal sacrifice is made on both sides to seal, as it were, their friendship, and the alliance just made between them (cf. 511-19 above). It also follows on the sign in heaven sent by Venus (520-31).

546 **graditur** - *Sc.* Aeneas.

socios - Aeneas' men who accompanied him on the voyage up the Tiber. It seems as if they passed the night in the boats.

547 **qui...sequantur** - Purpose clause introduced by a relative: 'who are to...' or 'to follow...'. The corresponding English idiom is expressed by the infinitive.

548 **prona** agrees with *aqua* in the following line: although the last syllable of a hexameter may be short, it is clear that *prona* describes *aqua* and not *pars*.

549 **segnis** - Adjective with an adverbial meaning: not 'idly' or 'indolently' so much as 'following an easy way'. The current (*secundo...amni*) would carry them easily downstream without any need for rowing. Aeneas has chosen the best men to stay with him, but only a few would in any case be needed to steer the boats back. Strong rowers have become superfluous.

550 **nuntia** - Adjective agreeing with *pars* (548).

ventura - From *venio*.

rerumque patrisque - Genitive after *nuntia*.

551 **dantur equi** - Evander provides horses for those who are to ride to Agylla to obtain help from the Etruscans under Tarcho.

petentibus - The present participle anticipates the future: they are so eager to set out that they seem to be already on the way.

552 **ducunt** - 'they bring out...'.

exsortem (*equum*). Aeneas is given an especially fine horse.

552-3 **quem...aureis** - A touch of pageantry after Vergil's love of the colourful and picturesque. The horse-cloth gives a bright touch to the group of horsemen as they ride away with its tawny skin and gilded claws as they glitter on Aeneas' horse.

aureis - In the emphatic position, drawing attention to the glitter. This word is scanned with two syllables, $\overline{au}r\overline{e}is$.

NOTES (VIII, 554–60)

554–5 **Fama volat** - This expression, being equal in meaning to a verb of speaking, governs the accusative and infinitive construction, *ire equites*....

555 **ocius** - Comparative form, but often used to mean 'quickly' as here. There may be an intensified meaning, 'very quickly', to heighten the feeling of excitement and tension which is general at this moment.

556–7 **propius...timor** - Fear comes nearer because of the growing danger, as war becomes imminent. Feelings are brought to a head when they see the Trojans setting off for Agylla.

556 **periclo** - Ablative of cause.

557 **maior...imago** - 'the image (their thoughts, or conception) of Mars (the war) looms greater (more menacing)'. These lines contain an analysis of the process of growing fear which illustrates Vergil's deep understanding of the human spirit. Here sympathy is shown for the women and their anxiety.

558 **euntis**, from *eo*, refers to Pallas. Evander clings to his only son while he bids farewell.

559 **inexpletus lacrimans** - He cannot stop his tears.

560–83 Evander's words of grief-stricken farewell. Full of the despair which comes from old age he thinks of his own achievements when he was young. His grief mounts, expressed more intensely by alliterative sounds indicative of woe: cf. 576 where the repetition of *v* sounds seems to imitate weeping, and even is continued into the next line: and *v* sounds occur frequently from 571 to 579, being more numerous from 576 to 579, to add to the sense of lamentation and despair: the climax of this effect of sound is perhaps to be recognized in the repetition of the heavy monosyllable *dum* (580–1), which seems to complete Evander's foreboding of the tragedy to come. His words end abruptly when he imagines how it will be if some message comes that Pallas has fallen. He can go no further and suddenly breaks off, unable even to speak final words of goodbye. This ending, more dramatic than a torrent of words, touches the reader's heart deeply, and leaves a silent anguish to be keenly felt. This speech should be read through again and again to appreciate to the full the pathos and sympathetic understanding of old age.

560 **referat** - Subjunctive expressing a wish: *o* and *si* = *utinam*.

87

NOTES (VIII, 560-4)

The present tense is used for vividness. The wish, referring to what is past, cannot be granted, but in Evander's desire it becomes almost a reality. With this is associated the underlying thought, 'and if Jupiter were to restore my youth...', to which *divellerer* (568), *dedisset* (570), *viduasset* (571) form the apodosis.

561 **qualis eram** - In loose apposition to the preceding line: Evander wishes that he could be the man he once was.

primam - Either the first exploit which Evander performed, or the front rank of battle: the second meaning makes Evander's exploit all the greater.

Praeneste - Usually neuter, but here feminine.*

562 **stravi** - From *sterno*.

incendi - The practice of burning the shields of the conquered was thought to have been introduced by the Roman king Tarquinius Priscus who burned the shields of conquered Sabines in honour of Vulcan. The custom continued to be observed from that time. Vergil gives us here a touch of history belonging to a later age than the legendary period of the *Aeneid*. There are many instances in the poem of such projections of history into the legend, for Vergil loved to touch on history in his writing.

563 **hac** - While Evander speaks he clings to Pallas and holds tightly his right hand (558); cf. 567. As he speaks, he thinks of the former strength of his own right hand and what it had accomplished.

sub Tartara - A way of saying 'to death'. Tartarus was one of the regions of Hades, the shadowy underworld where the dead were thought to go, according to the current Roman thought of Vergil's time.

564 **cui** comes, in meaning, in front of *nascenti*. Erulus was born with three lives.

animas means either that Erulus was a warrior with three lives to lose and that each time he was apparently slain he rose up again fully armed, or that he had three bodies. In this latter form he can be compared with the giant Geryon who was slain by Hercules in his Tenth Labour. This is not however mentioned in the hymn sung by the Salii in praise of Hercules (293-302). Each of Erulus' three bodies was fully armed as a warrior, causing Evander to fight three separate combats before he was finally slain. There is a purposeful

88

NOTES (VIII, 564–71)

connection to be found between the former strength which Evander recalls and that of Hercules. For this reason, perhaps, Hercules did not despise Evander's poor home, for he recognized a strong man who could match him in valour, in combats with monsters (185–275, 362–5 above).

Feronia - A goddess connected with springs.

565 **dictu** - Ablative of the supine with an adjective to denote the point to which it has reference: 'in the telling', i.e. 'to tell'.

566 **leto** - Ablative, 'in death'.

sternendus - Gerundive expressing what had to be done.

cui - Dative of disadvantage, used with verbs of depriving, taking away, etc.; with *abstulit*, *exuit* in next line.

567 **armis** - Ablative of separation, 'stripped him from his arms'.

totidem, i.e. 'three times', corresponds to *ter* (566).

568 **divellerer** - As the apodosis to *referat* (560) the present sequence would be expected, but since the wish is quite hopeless which is expressed in *referat*, the imperfect is used in *divellerer* in conformity with the underlying thought. The imperfect tense gives a pictorial, living effect which increases the pathos of Evander's last farewell: 'I should not now be being torn...': the continuous verb emphasizes his agony.

568–9 The endings *usquam*, *umquam* are lingering, sorrow-filled words. They sound the depths of Evander's despair and fruitless longing.

569 **tuo** agrees with *amplexu* in the preceding line.

569–70 **finitimo...huic capiti** - *finitimo* agrees with *mihi* understood, although some authorities take it with *huic capiti*; but regarded in this way it loses force. Evander is grieved that one who is his neighbour can insult him—cause so many of his own people to die (in battle)—and at the same time that he has not been able to help the neighbouring people against the tyrant. He first speaks of himself as Mezentius' neighbour (*finitimo*) and then of his own self (*huic capiti*). *Caput* is often used to mean 'life' and then 'an individual'.

570 **dedisset** - Pluperfect subjunctive, although in the apodosis to *referat* (560), because it expresses what cannot be; cf. *viduasset* (571).

571 **viduasset** - Shortened pluperfect subjunctive.

civibus - The ablative is regular with this verb; probably one of separation.

urbem - Either Agylla which had suffered from Mezentius'

NOTES (VIII, 571–82)

cruelty, or Pallanteum which is even now being deprived of its fighting men. Evander anticipates the future which will bring more sorrow.

572 Evander addresses the gods; the second part of his farewell becomes an impassioned prayer that he may live if Pallas is to be spared: but if Fortune decrees disaster, may he die before he can know of it. He speaks now as a father pleading for his son, not as a mighty man of strength.

maxime - Vocative; Jupiter was often addressed as *optimus maximus*.

573 **regis** - Genitive, regularly used with verbs of pitying.

574 **patrias** - From adjective *patrius*.

576 'If to see him (again) I live and to meet him (again).' The future participle is frequently used with the meaning 'likely to...'.

unum - Neuter: 'to come at one with him'.

577 **patior...durare** - Unusual use of *patior* followed by a prolative infinitive (i.e. one which completes the meaning), but the two words carry almost the same meaning. They emphasize, by repetition, what Evander will have to face. Although he is wearied with age (*obsitus aevo*, 307), he is ready to endure old age longer still if only he may see his son again.

laborem - The toil of old age which has become a burden.

578 **aliquem infandum casum** - Such is Evander's fear that he cannot bring himself to name it.

579 **crudelem** - Life would be cruel if Evander were to outlive Pallas: he prays that he may die now before ever such news, if it has to be, can ever reach him.

abrumpere - Life is thought of as a thread which can be snapped off.

580 The words in this line are closely condensed: they express love, anxiety, hope, fear, foreboding: 'while my anxiety (for you) still has cause for doubt, while hope for the future still remains unsure'. Evander wishes to live so long as there is a chance that Pallas may return.

dum - Supply *sunt* and *est*: implies 'as long as...'.

581 **mea...voluptas** - Pallas was Evander's only child and was born to him late in life. This emphasizes the old man's age all the more and his great love for Pallas.

582 **neu**=*et ne*, and introduces a negative with 'and may not

NOTES (VIII, 582–91)

tidings too hard (to bear) wound my ears'. Again Evander dare not say what is really in his heart: even to mention death might be an evil omen.

583 **supremo** - A word often used for the last farewell to the dead; its use here prepares us for the slaying of Pallas, although all through Evander's words there is all too apparent his own foreboding, and to us the inevitability of the coming tragedy is made clear.

584 **fundebat...ferebant** - Imperfects used to give a living picture: Evander swooned as he spoke and was carried away.

conlapsum agrees with Evander: from *conlabor*.

585–8 A glittering procession is described after Vergil's pictorial style: it is beautiful with the glitter of arms, touches of colour, and enhanced by the figure, glorious and pathetic, of the young Pallas, so soon to fall a victim to war.

585 **adeo** - Carries emphasis and marks the change of scene: 'mark you', or 'look'. Attention is centred now on the departure of Aeneas and Pallas for Agylla.

exierat - The pluperfect marks a kind of finality.

587 **inde** - Next in the passing column of horsemen.

587–8 **agmine...medio** - It was customary for young soldiers to ride inside the ranks so that the tried men could protect them. This is true perhaps for Pallas, but it is also a place of honour.

588 **chlamyde** - A crimson cloak was worn by Roman generals: a foreshadowing of a Roman custom.

pictis...armis - 'blazoned arms', either painted or inlaid with precious metals.

conspectus - 'looked at (on every side)'. Perfect participle with the meaning of *conspicuus*. All eyes are on Pallas as he rides.

589–91 Pallas is as bright and fresh as the newly-risen morning star: he is in the brightness of youth and loved as a son.

589 **Lucifer** - The morning-star, the name for the planet Venus when it rises in the morning: at its evening rising it was called *Hesperus* (the evening-star). This passage is one of Vergil's most beautiful similes (cf. x, 454–6 and note).

591 **extulit** - Gnomic perfect describing what regularly happens at any time, past, present or future. Translate by a present tense in English idiom.

NOTES (VIII, 592–8)

592 **muris** - The women of Pallanteum watch them go from the walls of the town with dismay in their hearts at the thought of war.

593 **nubem** - In a hot country such as Italy and at that time of the year (August) all grass is dried up to stubble, so that any movement of animals, especially galloping horses, raises a cloud of dust. Their arms and armour however glitter through it.

594 **olli** - Archaic form of *illi* used to give an effect of grandeur.

dumos emphasizes how the site of Rome is grass-grown. They go over rough ground, for there is no other, and follow rough tracks.

qua...viarum - 'where is the shortest goal of the ways', i.e. they follow the shortest path to their goal. Their way would be first back along the track, later the *Via Sacra*, which led through the valley where the future Roman Forum was to be, down to the Tiber. Here they would cross either by fording the river, which might have been possible at that time of the year when the water is very low, or by a primitive bridge. Vergil may have thought that the oldest bridge in Rome, the *pons Sublicius*, was older than the foundation of the City. Just how he visualized their crossing must be left to conjecture. Once on the far side they would follow an old road, then only a track, to Agylla.

596 A most remarkable line for imitation of sense in the sound of the words. The hurried movement of five dactyls gives the effect of galloping hoofs; muted *-em* and *-um* endings, together with hard *d* and *t* sounds, are reminiscent of the beat of hoofs on the dry ground. Read this line over many times: it is perhaps unparalleled in the whole of the *Aeneid* for sheer mastery over words.

597 **prope** comes in meaning before *gelidum...amnem*. This stream was one of several which flowed near Agylla: most probably that on the south, nearest Rome, now called the 'Vaccina'.

Caeritis - Genitive of *Caere*, the other name for Agylla. This place, now occupied by a village called Cerveteri, was one of the wealthiest Etruscan cities, situated some twenty miles to the north of Rome.*

598 **patrum** - '...practised by our forefathers'; subjective

NOTES (VIII, 597–605)

genitive which stands in the same relation to a noun as does a subject to a verb.

597–9 A pleasant description of the scenery near Agylla. The sanctuary (*lucus*) is beside a stream, in a hollow among the wooded hills. This is as true today as it was in Vergil's own time, although the site of the sanctuary is not now known.

600 Silvano - *Silvanus* was a god of the wild, belonging to the uncultivated land outside the boundaries of a town or village. He was also connected with forests and presided over flocks and herds since they were pastured in the uncultivated country outside the town (601). The camping-place which Aeneas chooses near this country shrine is outside the limits of Agylla. In this way he avoids violating the territory of the Etruscans who live there.

fama est governs the accusative and infinitive construction (*sacrasse Pelasgos*).

sacrasse - Shortened perfect infinitive.

Pelasgos - A name often applied to the earliest inhabitants of Greece who were thought to have settled also in Italy. Vergil uses it here for the inhabitants of Agylla who were believed to have come from the East (479). The name, derived from the Greek, appears to mean 'sea-men'.

601 deo - In apposition to *Silvano* in preceding line.

lucumque diemque - The worship of Silvanus had been established by the Greeks when they first settled at Agylla. A clearing in a grove was the characteristic place for the worship of a deity of the countryside. Here there would be a small altar. Every deity had a festal day on which special ritual was observed (*diem*) and a holiday kept. Silvanus was often identified with the Greek god of the countryside, Pan.

602 fines...Latinos implies the lands which are now inhabited by Latins.

603 tuta goes closely with *locis* in the next line; implies a place of natural defence.

603–5 Tarcho is in command at Agylla now that Mezentius has fled to Turnus at Ardea. He has pitched a camp for the revolting troops, which he leads until the destined leader is found, on a plateau on which they can be seen by the advancing Trojans because their tents are conspicuous.

605 legio - Tarcho's troops.

607 **succedunt** - The preposition *sub*, with which this verb is compounded, implies that they make their way 'up to'...; cf. 604 *celso...de colle*.

corpora curant - Lit. 'they refresh their bodies', i.e. 'they take rest and food'; a regular phrase in military language.

Aeneid x, 362–79, 439–509

362–8 A long, involved sentence, the main clause of which is ...*Pallas...virtutem accendit*. When translating, bear this in mind, but keep the Latin order of the subordinate clauses (four in all): *qua...*, *ut vidit...*, (*aspera quis...quando...*), *unum quod....*

362–3 The Arcadians are obliged to follow the enemy down the bed of a sunken stream typical of many which drain the Roman Campagna (ancient Latium) today. They have ridden down the edge of the Tiber from Pallanteum.

362 **parte ex alia** = *ex alia parte*, 'in another direction'.

saxa rotantia - The participle is used intransitively. The boulders have been rolled about by the stream.

363 **ripis** - Ablative of separation.

364 **Arcadas** - Greek accusative: note *insuetos* in agreement with it.

insuetos - The Arcadians were essentially horsemen: the rough nature of the ground they have reached means that they must lead their horses: the result was that unused to fighting on foot they were losing heart. In the Roman army it was customary in difficult country to hand the horses over to grooms (*comites*) and to fight on foot.

364–5 **inferre** comes in meaning after *insuetos*, and *dare* after *vidit*.

365 **Latio** - 'the Latins'. The name of the land is used for the people: the Arcadians were giving way before the pursuing Latins.

366 **quis** - Dative plural of *qui*, more usually *quibus*, with *suasit* in the next line.

quando - Equal in meaning either to *quando-quidem*, 'since', or more probably to *ali-quando*, 'for once'.

367 **unum** looks on to the next line: their one hope of safety lies in rallying to make a stand on foot.

369–78 Pallas, the young inexperienced warrior taking on the

NOTES (X, 369–78)

responsibilities of leadership, addresses his troops just as a Roman general was accustomed to do before battle.

369 **fugitis** - A word full of reproach, and corresponding to *amaris* in the preceding line.

fortia facta - Pallas reminds them of their own former achievements: he has none of his own as yet because he is untried in war. His exploits are about to come.

370 **ducis Euandri** - There is a special significance in calling their king their leader: it reminds them of his former strength, and the example they have to follow (VIII, 560-7).

devictaque bella - 'wars fought and won'. The preposition *de*, when compounded with a verb, usually denotes an action which is carried out to the end.

371 **spem meam** - Pallas' hope of doing some glorious deed on the field: 'my (own) hope which now comes up (to my mind) (as) rival to my father's glory'. Pallas has a sudden impulse to try and match his father's achievements.

372 **fidite ne**=*ne fidite* (note on VIII, 532).

pedibus - Dative with *fidite*. They must not rely on flight, but must face the enemy. They must cleave a way for themselves through the enemy's lines.

373-4 **qua** is answered by *hac* in the next line. *viā* (ablative) is to be understood.

373 **ille** - Deictic. Pallas points to the direction in which they should go.

374 **alta** - 'noble'.

375-6 Pallas reminds his comrades that the Latins are only men like themselves: they are not fighting against monsters as Evander once did when he slew Erulus (VIII, 560-7), who had to be overcome three times as he had three lives and three bodies to Evander's one.

377 The scene of the fighting is the shore of Latium: if they flee they will reach the sea and there they can find no help.

magna...obice - The feminine is an archaism: the ablative is instrumental. The repetition in meaning in *maris, pontus* suggests that the watery barrier in their way is insuperable.

378 There is no safe direction for flight on land.

fugae - Dative.

Troiamne - I.e. New Troy: Aeneas' camp on the bank of the Tiber. Flight here is impossible for the Arcadians, because the enemy surround it.*

NOTES (X, 379–444)

379 **medius** - Used proleptically. Pallas himself plunges into the midst of the enemy, hoping that the others will follow.

439 **soror** - Turnus' sister Juturna, an Italian nymph associated with healing waters: her name is probably connected with *iuvo* ('help') and in her character as a goddess of healing she is called *alma*, 'fostering'. In the Forum in Rome there was a spring sacred to her at the foot of the Palatine Hill. This, which was in the form of a well, was decorated with masonry and sculpture in Roman times, and regarded with reverence. It was said that the twin gods Castor and Pollux miraculously appeared there after fighting at the side of Postumius, the Roman dictator, at the battle of Lake Regillus (499 B.C.). They brought the news of victory to the waiting populace, watered their tired horses in Juturna's spring and then were not seen again. A temple in their honour was built beside the spring; the three remaining columns still standing on a high podium are among the most beautiful of any remains in the Roman Forum to be seen today. Vergil brings Juturna into the *Aeneid* as the deified sister of Turnus.

succedere Lauso - 'to help' in the sense of 'taking his place'. Lausus the son of Mezentius, fighting on Turnus' side, is weary. He is the equivalent of the young Pallas on the Trojan side. They are not destined to meet each other in combat, although each has his deeds of heroism, and neither will return home. In prose *ut* with the subjunctive would be expected with *monet*.

441 **ut vidit** - *ut* is used with the indicative with the meaning 'as' or 'when'.

tempus - Supply *est*.

442 **feror in** - 'I am borne against', i.e. 'I confront', with a sense of obligation; it is Turnus' inescapable destiny to fight with Pallas.

443 **cuperem** - 'I could wish that his father himself were here to see (his end)'. Such a thought is in keeping with Turnus' savage, bloodthirsty disposition. To slay a son before the eyes of his father was considered to be an act of supreme wickedness. In *Aen.* II, 526–32, Pyrrhus slays Polites before the eyes of Priam, his father. The imperfect subjunctive is used to denote a wish which is not possible of fulfilment.

444 **socii** - Turnus' comrades withdraw to leave an open space

NOTES (X, 444–53)

for the combat with Pallas. *aequore iusso* = 'when an open space had been ordered'.

445 **Rutulum** - Genitive plural.
iuvenis - Pallas.

446 **stupet** - Pallas is amazed and awed now that he is able to gaze at Turnus, seeing him for the first time face to face.

447 **omnia** - Turnus' body and arms. Pallas looks him all over and is not afraid, for he looks *truci visu* and then defies him.

448 **it contra dicta** should mean 'he answers', but Turnus has not yet spoken to him. Pallas speaks in answer to Turnus' command to his men to stand aside (441-4).
tyranni - Turnus.

449 **spoliis...opimis** - Another anticipation of Roman customs. The *spolia opima* were the trophies won by a Roman general from the general of the enemy, whom he had slain in single combat. It is said that this happened only three times. They were won by Romulus in legendary times, and in historical times by Aulus Cornelius Cossus when he slew Tolumnius King of Veii (428 B.C.) and by Marcus Claudius Marcellus when he won the victory over Viridomarus at Syracuse (221 B.C.). The *spolia* were dedicated in the temple of Jupiter Feretrius in Rome. This temple, thought to be the first ever built in Rome, was situated on the Capitoline Hill, and was said to have been built by Romulus to commemorate his winning the *spolia opima* from Acron king of the Caeninenses, and to serve as a place in which to dedicate them. Pallas' desire to distinguish himself in this manner shows him to be endowed with Roman ideals of heroism of the time to come.

450 **sorti** - 'my father is equal to either fortune', Evander would resign himself to either glory or disaster: an answer to Turnus' base desire that Evander might be present (443). In this attitude we see a typical Roman father.

452 'Cold gathered the blood in the Arcadians' hearts.'
Arcadibus - Possessive dative. Their blood runs cold and seems to congeal in their hearts: their fear and foreboding is realistically expressed: they know that Pallas cannot match the tried warrior Turnus.

453 **pedes** - Adjective agreeing with Turnus. The fight is to be on foot. The fact that the Arcadians are really horsemen should not be forgotten nor the fact that Pallas rallied them

NOTES (X, 453-61)

to fight on foot. The combat becomes all the more poignant when we consider Pallas' ordeal in the light of this fact.

454-6 A simile taken from the animal world. Similes are often introduced into the *Aeneid* (VIII, 589-91 and note) to heighten an effect, to give pause before some great dramatic moment, to change a scene, to bring in more vividness. Here the tension is increased. We look for the outcome but are held back by another image. Turnus is like a lion falling on its prey: his speed is consonant with his nature: *advolat* alone in 456 seems to suggest that he does not hesitate.

456 **imago** - 'the sight of...' or 'appearance of...'. Turnus is like a lion rushing on its prey.

457 **hunc** - Turnus.

contiguum...hastae - 'within the throw of a spear'. As Turnus dashes towards him, Pallas judges the moment when he is within throwing distance.

fore - Future infinitive of *sum* (*futurum esse*).

458 **ire** - Historic infinitive.

prior - Pallas is the first to strike.

458-9 '(to discover) if in any way fortune may help him daring with unmatched strength (i.e. although his strength is less)'.

adiuvet - Subjunctive in an indirect question: the word sounds a note of despair.

ausum - From the semi-deponent *audere*: agrees with *se* (i.e. Pallas) understood. Some scholars however take it to be a neuter noun, the object of *adiuvet*, 'his enterprise': in this interpretation *viribus imparibus* (459) goes closely with *ire prior*.

459 **aethera** - Greek accusative: *magnum* agrees with it.

460 **per...hospitium** - It is wholly appropriate that Pallas should pray for help to Hercules, for not only did he visit Evander's home at Pallanteum and stay there as his guest after slaying the monster Cacus (VIII, 185-275), but also the Arcadians, according to Vergil, established the cult of Hercules at the *ara maxima* on the bank of the Tiber in recognition of his great services (VIII, 185-305).

461 **Alcide** - Vocative of the patronymic, 'descendant of Alcaeus': one of the names of Hercules (for another see VIII, 103 and note). **coeptis ingentibus** - Dative with *adsis*, 'my mighty beginnings', i.e. 'my mighty aspiration'. Pallas has set himself a great deed of daring, to attack Turnus who is an experienced warrior, older and stronger than himself. It is

NOTES (X, 461–71)

called a 'beginning' because it is as yet his ambition, it has not been accomplished.

adsis – Present subjunctive in a prayer, dependent on *te precor*: this construction is normal without a conjunction as here: cf. *cernat* (462) and *ferant* (463).

462–3 'May he (Turnus) see me (when he is) half-dead stripping from him his bloody arms, and may Turnus' dying eyes endure me (i.e. the sight of me) as the victor.' This expressed hope is in answer to Turnus' wish that Evander might see his son slain. It seems as if in words at any rate cruelty must be matched with cruelty, although Pallas does not go to the same lengths as does Turnus.

462 **sibi** – Dative of disadvantage with verb of depriving.

465 **lacrimas** – Hercules' tears were of no avail because it was not given to Pallas to be victorious. There was an unchangeable destiny stronger even than the desire of the gods.

466 **genitor** – Jupiter (VIII, 103 and note).

amicis – Used as an adjective.

467–72 Jupiter's answer contains a peculiarly Roman view of life. Each man has an allotted span of life which cannot be altered. He should therefore aim to make his life distinguished by great deeds or achievements so that it may continue in the memory of these. The wish was expressed by many Latin writers including Ennius, Horace, and the Younger Pliny.

467–8 'Stands (fixed) his own day for each (man): brief and unchangeable is the span of life for all.'

sua agrees with the subject *dies*.

cuique – Possessive dative.

dies – In the feminine gender because the meaning is 'length of time', not merely 'a day'; in the latter meaning the word is always masculine.

468 **extendere** – 'to lengthen in time'.

469 **opus** – 'task', i.e. that which needs to be done.

470 **cecidere** – From *cado*.

deum – Jupiter reminds Hercules that even the sons of gods fell at Troy, and even his own son, Sarpedon (471).

quin – An exclamation leading up to the climax that even Jupiter's own son fell at Troy.

471 **Sarpedon** – Commander of the Lycians who fought for Priam. He was prominent in the fight, and made the first

NOTES (X, 471-86)

breach in the Greek wall. In the end he was killed by Patroclus and mourned by Jupiter (Zeus in the Greek). Sleep and Death took him back to his native land, Lycia, for burial (*Iliad* XVI, 477 ff.).

472 **metasque** - Turnus' life is nearly accomplished. There is perhaps comfort in the thought that he will not long survive Pallas.

473 Jupiter watches the fight no longer, because he cannot intervene.

475 Pallas draws his sword after throwing his spear, to be ready if Turnus attacks him close at hand.

476 **umeri** - 'where the topmost edge of the coat of mail rises on (lit. "of") the shoulder'. *tegmina* literally means 'coverings'.

477 **molita** - '(the spear) forcing a way through the rim of (his) shield'.

478 **strinxit de corpore** - '(just) grazed his body'. The spear made a slight scratch. *de* can be compared to the partitive genitive in Greek giving the sphere within which an action takes place.

480 **diu** heightens the tension with a dramatic pause while Turnus' counter-blow is awaited.

481 **mage** - With *penetrabile*, to form the comparative.

482-5 The construction of this sentence is: *clipeum* (object)... *cuspis* (subject)...*transverberat*,...*loricaeque moras et pectus* (object) *perforat*. The relative clause *quem...tauri* includes in meaning also *tot ferri terga, tot aeris*. As far as possible keep the Latin order in translation: 'but the shield which though so many plates of iron..., yet (this) his spear-point...struck and pierced...'.

482 **clipeum** - Ancient shields were made on a framework of wicker attached to a metal boss of iron or bronze. Layers of leather were attached to this which were bound at the edge with metal.

483 **obeat** - Subjunctive with the relative *quem* (483), carrying the suggestion 'although', i.e. 'in spite of the strong covering of the shield'.

485 **loricae moras** - 'the defences (causing delay) of his cuirass'.

486 **rapit** - Pallas pulls out the spear, but in vain because he is mortally wounded.

NOTES (x, 487–97)

487 eadem - Ablative: scan as two syllables.

sanguis - The lengthening of the last syllable (*sanguīs*) is remarkable. The stress falls on the key word in the line.

490 One of the most dramatic short lines in the poem. There is a pause which prepares for a worse horror than Pallas' death, the violation of the body (cf. VIII, 41 and note).

492 qualem - I.e. Pallas. Turnus says that this is the repayment to Evander for sending his son against the Latins. A just requital in Turnus' eyes.

493 humandi - Genitive of the gerund used intransitively: common in Vergil.

494-5 'At no small cost to him (Evander) will stand the hospitality given to Aeneas.' Turnus has won so great a victory that it will cost him nothing to give up the body of Pallas.

494 parvo - Ablative of price, commonly used with *stare* and *constare* ('cost').

Aeneia for purposes of scansion is a four-syllable word.

495 Turnus heaps cruelty on Pallas by violating the body.

496 baltei - Scan as a disyllable (*baltēī*).

496-7 'Snatching the great weight of his sword-belt and the crime engraved on it.' Almost a hendiadys: the belt was a heavy one and had mythical scenes engraved on the metal facings.

497 The belt was engraved with the history of the Danaids and seems to stress the greatest horror of it, unless it is to be understood that various scenes were engraved on separate metal plates. The story, which belongs to a cycle of Greek mythology, told how Aegyptus and Danaus, being brothers, both claimed the kingship of Memphis. At last Danaus yielded to Aegyptus and travelled to the Peloponnese (southern Greece). There he became king and called his subjects after him Danai. To make peace between them Aegyptus suggested that Danaus' fifty daughters should marry his sons. Though his exile still rankled, Danaus consented, but, on the night of the marriage, at his instigation his daughters slew their husbands, all except one, Hypermestra, who spared her husband Lynceus. The Danaids then suffered the punishment in Hades of having to fill leaking water jars, a type of punishment which is never finished. This kind of myth would easily divide into separate scenes. This is probably how the belt was designed although Vergil does not give all the details, in

NOTES (X, 497–XI, 25)

accordance with his usual manner of selecting rather than giving all. The mention of the artist, otherwise unknown, marks the belt as a masterpiece of the metalworker's art.

498 **caesa manus** - We should expect these words to be in the accusative since they are in apposition to *nefas* (497), but the nominative is much more expressive: *caesa* seems to leap out at us in all its horror with the sudden nominative statement.

499 **auro** - The metal plates are inlaid with gold.

500–2 The belt will be Turnus' own undoing. A philosophic reflection on the weakness of human nature. Men do not know how to restrain themselves in times of success.

501 **fati, sortis** - Objective genitives with *nescia*.

502 **servare** - Prolative infinitive depending on *nescia*, an unusual usage.

503 **Turno** - Ethic dative.
magno - Ablative of price, recalling his own words (494) spoken tauntingly of Evander.
optaverit - Future perfect because when that hour comes his wish will be ended (in death). This is a vivid use which can hardly be fully expressed in English.

506 **impositum...referunt** = *imponunt et referunt*; *impositum* is used especially of burying the dead.
scuto - A very pathetic touch because the shield bears the marks of his blood where Turnus' spear pierced it.

507–9 Addressed to Pallas.

507 **rediture** - Vocative of the future participle.

509 **tamen** - Used in a peculiarly Vergilian manner. An unexpressed thought must be supplied; such as that, in spite of all the grief and pity, yet Pallas has had his triumphs: he has slain many *Rutuli* all in one day and has achieved his *gesta* (exploits) just as a seasoned warrior might. He does not go home without distinction.

Aeneid XI, 24–99, 139–81

24–8 Aeneas' last commands to his men.

24 **animas** - Object of *decorate* (25).

25 **patriam** - Aeneas looks to the future: 'this new fatherland'.
peperere - Perfect, from *pario*.

NOTES (XI, 25–34)

25–6 supremis muneribus - The last rites which had to be observed at the burial of the dead (see below, 29–41 and notes).

26 primus agrees with *Pallas* (27).

27 mittatur - Subjunctive expressing a command (jussive).

non...egentem - Litotes, implying 'full of bravery'.

virtutis - The genitive as well as the ablative is commonly used with *egeo*, to denote the sphere in which the want is felt.

28 atra - *Ater* in origin means 'black'; derived meanings are 'gloomy', 'dark', 'foreboding', 'ill-omened'. The Romans used to mark unlucky days in the calendar with a black mark: hence *dies atrae* were days on which calamities had occurred and were always remembered as being unfortunate.

29 limina - The threshold of Aeneas' quarters in the New Troy.*

30–1 Take in this order: *ubi Acoetes senior exanimi Pallantis corpus positum servabat*. Turnus has allowed Pallas' body to be taken within the Trojan camp (x, 492).

30 positum - 'laid (on the bier)'. It was customary among the Romans for a dead body to be laid out on a bier in the hall (*atrium*) of the house with the feet pointing towards the door: cf. the mention of *limina* (29) and *fores* (36).

31 servabat - 'was watching over'.

senior - The force of the comparative has been lost in the use here; it is merely 'old'.

Parrhasio Euandro - Notice the unusual scansion: there is a hiatus between these words, and a spondee in the fifth foot (spondaic ending). The resultant effect is one of grief, almost suggesting sobbing and catching of breath, emphasized by the long *o* sounds.

32–3 Take in this order: *sed non felicibus aeque auspiciis tum comes caro alumno datus ibat*. Acoetes was not as fortunate in his service as when he had been Evander's squire.

33 alumno - I.e. Pallas. It was customary for a father to put a young warrior in the charge of an older retainer, or 'squire', who would attend him and watch over his safety and welfare.

34–5 The mourners stand around the couch on which the dead body of Pallas is laid. After the Roman custom they carry out the wailing for the dead (*conclamatio*).

34 famulum=*famulorum*.

NOTES (XI, 35-42)

35 **solutae** - From *solvo*: for the construction of *crinem solutae* see VIII, 286 and note. To wear the hair loose was a sign of mourning. The women of Troy who had sailed with Aeneas and his men and remained with them in the camp take part in the mourning.

36 **foribus** - Dative for 'motion to', frequent in Vergil. For the mention of *fores* see note on 29 above.

intulit - From *infero*.

37-8 **tunsis...pectoribus** - 'as they beat their breasts'. This action was a sign of mourning in ancient times and accompanied the lament: *tunsis*, although a past participle, describes a continuous action in this context—the sound of wailing increases on Aeneas' return. These two lines are remarkable for onomatopoeia (sound imitating sense); notice especially *gemitum, maesto, immugit, luctu*, all sounds of grief and lamentation.

38 **regia** - Usually the word for a king's house, but here for Aeneas' quarters in the camp: perhaps only a lowly hut, but receiving a royal epithet in acknowledgement of him as a prince, for his father Anchises belonged to the younger branch of the royal house of Troy.

39-41 The construction of this sentence is: *ut...caput... Pallantis...que...vulnus cuspidis...vidit...ita fatur*. These lines express the depths of human pathos. What Aeneas sees first, and continues to gaze on, is Pallas' head, deathly-white resting on the pillow, and the deadly wound: two focal points which bring out all the more the tragedy of his boyhood sacrifice.

40 **levi** - From *lēvis* as scansion shows. This adjective emphasizes the beauty of his young, tender body and the cruelty of the gaping wound.

41 **obortis** - From *oborior*.

42-58 Aeneas' lament over the dead Pallas is surpassed only by that of Evander (152-81 below) for deep-felt emotion, for grieving at the death of an innocent victim of war, and the irony of a fate which might have been glorious.

42 **tene** - From *te* and *-ne*: in the emphatic position. The construction of the sentence is: *tene...invidit Fortuna...ne... videres...neque...veherere*, 'was it you, poor boy, whom Fortuna, when she came rejoicing, grudged me, to prevent your seeing...'.

NOTES (XI, 42–50)

laeta agrees with *Fortuna*; is almost adverbial in meaning.

43 **ne...videres** - Negative clause of purpose.

43–4 **regna...nostra** - Aeneas' kingdom in Latium has yet to be established: he anticipates the future.

44 **neque** = *et* and *ne*, but *neu* or *neve* are more usual in this sense.

ad sedes...paternas - 'to your father's home', i.e. to Pallanteum.

victor - 'as victor'; agrees with the subject of *veherere*.

veherere = *vehereris*, imperfect subjunctive passive from *veho*.

44–7 Lines full of alliteration, especially in pairs of words, *victor veherere* (44), *promissa parenti* (45), *discedens dederam* (46), *cum...complexus* (46), and most notably in 47, *mitteret... magnum...metuensque moneret*, where the alliteration in *m* sounds is enhanced by the word *imperium*. This whole line is full of the muted sounds of a sobbing voice: cf. with this the *m* sounds in the line before, 46, which seem to lead on to the intensity of 47.

46 **complexus** agrees with Evander.

euntem - Present participle from *eo*.

47 **mitteret...moneret** - Subjunctives with *cum* (46) meaning 'when'.

in magnum imperium - 'to (win) great empire': notice this use of *in* to express the 'end in view', with the idea of striving towards an object.

metuens agrees with Evander, subject of *moneret*.

48 **acres esse viros** - Accusative and infinitive construction after *moneret* in the preceding line.

cum...gente - Also dependent on *moneret*: 'that the battle was with a hardy race': supply *esse* with *proelia*. *cum* in this line is a preposition governing the ablative (*gente*): an indirect tribute to the fighting powers of the Latins.

49–52 These lines show the contrast between Evander, who does not yet know of his son's fate and is perhaps still praying for his safety and success, and those who mourn for the dead Pallas: the emphatic pronouns, *ille* (49) and *nos* (51), point the difference.

50 **fors et vota facit** - 'he (Evander) may even be offering prayers...'.

cumulat - Evander may be making gifts and vows to the gods in return for Pallas' safety.

NOTES (XI, 51–99)

51–2 nil...debentem agrees with *iuvenem* (51), i.e. Pallas. He owes nothing to the gods above, that is, the gods of this world, because his life is fled, and he has gone to the realm of the dead: the vows made for his safe return have not been fulfilled, so now nothing remains to be paid. These words sound the finality of death.

52 vano...honore – Ablative of manner: usually with *cum* in prose.

53 infelix – Vocative, addressed to Evander. The sudden change to the second person intensifies Aeneas' expression of sorrow: he sees vividly the tragedy which Evander has shortly to face, and exclaims in anguish.

54 A line full of bitter irony: 'Is this what our triumphant return was to be, (and yet, see how it is)?' Supply *sunt*.

55 haec – Supply *est*: the question is as full of bitterness as that in 54.

at marks a change in thought. Aeneas has been speaking in tones of grief, but now he thinks with pride of Pallas' wound which is proof of the bravery he showed in facing Turnus.

pudendis – I.e. wounds in the back, which might have shown that he had fled.

55–7 'But you will not, Evander, look on (a son) who ran away (lit. routed) with shameful wounds, nor because your son is safe will you (his) father long for accursed death.' Evander has no cause to be ashamed of Pallas, for he has fought well and bears honourable scars.

56 sospite – Saved (by cowardly flight).

57 mihi – Ethic dative, 'ah me'.

57–8 Aeneas thinks of the loss to Italy and to his own cause.

58 Ausonia – A name for west-central Italy, but often applied loosely to the whole peninsula: supply *tu perdis*.

quantum – Supply *praesidium*.

Iule – Aeneas thinks of his own son, and in thought speaks to him.

59–99 The funeral procession which escorts the dead Pallas is typical of Roman customs. The corpse is laid on an open bier, covered with costly robes, and the outward signs of his achievements in life are displayed. As if it were a Roman triumphal procession, the spoils of war are on view—trophies of arms, captured horses and prisoners, and, bringing up the

NOTES (XI, 59–71)

rear, chariots spattered with the enemy's blood and Pallas' own war-horse Aethon.

59 **deflevit** - 'wept his fill'. *De* in a compound verb implies doing an action 'to the end'. *Defleo* is the usual word for mourning for the dead.

60 **imperat** governs *tolli* in the preceding line: poetical use of the infinitive with *impero*.
lectos - From *lego*, with *viros* in the next line. Aeneas sends a thousand chosen men from the Trojan, Etruscan and Arcadian contingents.

61 **qui...comitentur** - *Qui* with the subjunctive expressing purpose.
supremum...honorem - The last honour paid to the dead, i.e. the funeral procession.

62–3 **solacia...ingentis** - Note the contrast between *exigua* and *ingentis*. *solacia* is an accusative to be taken in apposition to the sentence regarded as a whole. Their escorting of the bier and their presence are meant to help Evander in his grief.

64–7 After the Roman custom, a couch (*torus*) is laid on the bier (*feretrum*) and a canopy (here described in the words *obtentu frondis*) set over it. All this is made in haste of such rustic materials as branches and leaves which are at hand. There is no opportunity for anything more elaborate. According to Roman ideas a costly bier, perhaps of polished wood ornamented with carved ivory and bronze fittings, would be provided for a person of distinction: the poverty and roughness of Pallas' bier and the couch laid on it serve to intensify the pathos of the mourning procession and of Evander's loss.

64 **haud segnes** - Litotes, 'not slowly', i.e. 'in haste'.
crates...feretrum - Hendiadys, 'wickerwork and a soft bier'='a soft bier of wickerwork'.

65 The bier is woven from branches and twigs of the wild strawberry tree and from the cork-oak, which are both common trees in the Italian countryside.

67 **sublimem** - 'on high', 'uplifted' in a literal sense.
stramine - *Stramen*, connected with *sterno*, means that which is laid down or strewn, often implying something to lie on: here it means the couch made of branches and leaves.

68–71 One of the finest similes in the *Aeneid*, expressed in words of exquisite delicacy and feeling. There is a lightness of touch which seems to match the youth of Pallas: he is as lovely as a newly-plucked flower left to die. It is a fitting ornament to the description of the funeral procession and to the dead boy couched on boughs and leaves.

68 **virgineo** - Perhaps one may think of a Roman garden like those found in the villas at Pompeii in which can still be traced laid-out flower-beds. In this simile, which gives us a glimpse of Roman life, it is a daughter of the household who plucks the flowers. There seems in this sense to be a contrast drawn between the young girl who is alive and the boy who lies dead.

demessum - From *demeto*: for the preposition *de* in a compound verb cf. note on *deflevit* (59).

pollice - A painting from Pompeii shows a girl plucking off flowers between her finger and thumb.

69 **violae, hyacinthi** - It is not certain exactly which flowers are meant by these names, but *violae* may reasonably be translated as 'violet', especially since this flower was often grown in Roman gardens. *Hyacinthus* is not the flower which we call 'hyacinth' now, but probably some species of iris. A story in Greek mythology concerning a flower called by this name told of a young and beautiful boy, Hyacinthus, who was loved by Apollo. When he was killed accidentally by a blow from a discus (some said that Zephyrus, who was jealous of Apollo's love for the boy, blew the discus out of its course so that it struck him on the head), a flower sprang up from the ground where his blood fell down. Its petals were marked with two Greek letters signifying 'alas!' The flower was forever afterwards given his name. Vergil introduces it into the simile because of its association with a boy slain in his youth: *languentis* recalls his death and at the same time suggests pallor and weakness. The rhythm of this line is imitated from the Greek and is found elsewhere in Vergil, as here, in connection with Greek words and names. Notice the lengthened *-is*, and the break in the fifth foot which is not otherwise allowable in the Latin hexameter. The effect of this exceptional metre is to slow down the pace of the line. The long drawn out syllables at the end seem to suggest the lingering swoon of death.

70 'Of which neither the bright colour nor as yet its fair shape

have withered.' The flower has only just been plucked and has as yet its full beauty.

sua - Notice this use of *suus* agreeing with the subject in the sense of 'own'.

71 The flower has been separated from the parent plant and no longer receives nourishment from the ground.

mater...tellus - In both Greek and Roman thought the earth was the mother of all things; in particular in this passage like a mother it nourishes and gives life to the plants.

72 **rigentes** - 'stiff with...'; the cloth was closely woven, and so of fine rich texture.

73 **illi** - Aeneas; dative with *fecerat* (75).

laeta laborum - 'happy in her toil'; she had delighted in weaving the material because it was for the man whom she loved. The genitive is one found with verbs which describe mental emotion, especially such as *taedet*, *piget*: it is probably an extension of the partitive genitive, but has also parallels in Greek. In giving up the two robes Aeneas relinquishes the most precious gifts he can find.

74 **Dido** - After long wanderings in the Mediterranean in search of a settled home, Aeneas reaches the shores of Carthage. He is received with kindness by the queen and lingers there for a year. Dido falls deeply in love with him and gives him many rich and costly gifts. When Aeneas leaves her, in despair she slays herself on a funeral pyre on the seashore. The story of Dido and Aeneas is told in *Aeneid* IV.

75 Dido had woven threads of gold into the material and had picked it out (*discreverat*) with the strands. *auro* comes at the end of the line to emphasize the richness of it.

76-7 The two robes are in form but two lengths of tapestry as they come from the loom. Aeneas folds one around the body and forms the other into a cloak which is put around the head and over the body.

76 **unam** - Supply *vestem* (cf. 72).

supremum...honorem - 'as a last honour'; *supremus* is a technical term for honours paid to the dead. *honorem* is in apposition either to *unam* (*vestem*) alone, or more probably to the sentence as a whole, as in line 62 above (see note).

77 **arsuras** - From *ardeo*: 'soon to be burned (on the funeral pyre)'.

NOTES (XI, 77–87)

amictu - The other *vestis* formed into a cloak.

78 **Laurentis...pugnae** - The fight against Turnus which took place in the Laurentian plain and near the city of Laurentum, King Latinus' city.

79 **duci** - Passive infinitive.

80 **equos** - Those captured in the fight. They would be burned, together with the arms and the prisoners, on the funeral pyre.

81–2 It was the Roman custom to make prisoners of war walk in triumphal processions and to put them to death immediately afterwards.

81 **et** comes in meaning before *vinxerat*.

quos mitteret - Relative with subjunctive expressing purpose (cf. 61 and 62 and note, above). The subject is Aeneas.

umbris - Dative with *mitteret*: 'to the shades of the dead', that is, to the shadowy underworld where the Romans believed the dead lived.

82 **inferias** - In apposition to *quos*: the prisoners would be slain 'as offerings' to the spirits of the dead.

sparsurus refers to Aeneas: he would cause their blood to be sprinkled.

caeso...sanguine - A Vergilian inversion for the 'blood of the slain'.

83 **truncos** - Tree-trunks, called *tropaea* (trophies), on which were hung arms captured from the enemy. These were to be carried in procession, but it was customary also to set them up in some place where they were conspicuous. This was often done on the field of battle. From this practice came the custom of carving trophies in stone.

84 **ipsos...duces** - An especial mark of honour, that the leaders themselves should carry the trophies and not assign the task to their soldiers.

nomina - The names of the opponents from whom the arms had been stripped were shown on the trophies (*truncos*, 83).

85 **ducitur** - 'is led along in procession'.

86 Beating the breast and scoring the cheeks with the fingernails were signs of mourning.

pugnis - From *pugnus*.

87 **sternitur** - As the *cortège* moves along, Acoetes at times falls prostrate on the ground, so great is his frenzy of grief, and he is an old man.

NOTES (XI, 87–141)

toto...corpore - 'full length': ablative of manner.

terrae - Dative, equal in meaning to *in terram*. A frequent usage in Vergil; some authorities suggest that this is, on the other hand, a survival of the old locative form.

88 **currus** - Chariots captured by Pallas and still bearing the marks of battle.

89 **positis insignibus** - 'his trappings removed', as a sign of mourning.

Aethon - From the Greek, meaning 'fiery'. One of Hector's horses in the *Iliad* had this name.

90 The idea that a horse can weep is also Homeric; but Shakespeare says of a wounded stag, *As You Like It*, II, i, 38–9,
> the big round drops
> Coursed one another down his innocent nose.

91 **cetera** - Especially the belt inlaid with gold which will cause before long his own death.

93 **versis...armis** - 'with arms reversed', i.e. pointing downwards, as a sign of mourning.

94 The whole army with Aeneas at its head escorts the procession a certain distance: then with the main body he leaves them and returns to the camp. The thousand picked men go all the way (61).

96–8 Aeneas pronounces the last farewell to the dead in words customarily used at Roman funerals, *salve* and *vale* (97, 98).

96 **alias...lacrimas** - The burial of others who had fallen in the fight.

97 **aeternum** - Neuter of an adjective used adverbially: it can be regarded as an internal accusative with *salve*.

mihi - Ethic dative, 'I bid (you)'.

98 **effatus** - From *effor*.

99 **tendebat, ferebat** - Imperfect tense in narrative.

139 **Fama** - A personification of rumour which flies fast, and is therefore thought of as a winged creature.

140 **Euandrum...replet** - 'fills Evander's mind': *replet* goes more easily with *domos et moenia* where the idea is logical that rumour fills a place: by an extension of meaning it is applied to a person.

141 Only just lately (*modo*) the news has been good: already the people in Pallanteum have heard of Pallas' triumphs on the field and have believed that he is victorious. They are ill prepared for the bad news which follows.

NOTES (XI, 141-55)

Latio - 'in Latium', on the Laurentian fields where the battle was fought.

142 **ruere** - Historic infinitive, describing here a quick-moving scene.

de more vetusto - It was the custom to carry torches in a funeral procession.

142-4 A most graphic description: a line of torch-bearers comes out from Pallanteum to meet the procession which approaches along a road over the open country. The two join to make one long line which shines across the fields. Since the distance covered from Aeneas' camp on the Tiber is something in the region of fifteen miles we may assume that by now night has fallen and that the fields are dark. Those escorting the dead body of Pallas have lighted their torches both in honour of the dead and to illumine the darkness.

145 **contra...veniens** - 'going to meet it'. They join the mourning ranks which escort the bier.

146 **succedere** - Infinitive with *viderunt* in the next line. *quae = agmina*.

147 **incendunt** - An unusual but startling metaphor: 'they kindle', or 'set the city on fire' with their cries.

148 **potis est** - An archaism for *potest*.

149 **in medios** - 'into their midst'. Evander cannot wait for the *cortège* to reach him, but he hurries through the throng of mourners to reach the bier.

feretro...reposto - Ablative absolute: as soon as the bier was set down, Evander flung himself upon Pallas: the ablative *Pallante* is governed by *super*.

151 The alliteration in *v* and *d* sounds gives a touching effect of the choking, broken sobs which hold back Evander's words.

153 **ut** = *utinam*, and *velles* following it expresses a wish: the imperfect subjunctive is used because it refers to past time and is therefore impossible of fulfilment. 'O that you had been ready to....' Some scholars connect *ut* with the previous line as expressing Evander's desire.

154-5 Evander well understands how the desire for glory in war can seize a young boy.

154 **nova gloria** - '(the desire for) new-found glory', i.e. the first taste of glory.

155 **posset** - Subjunctive in an indirect question, introduced by *quantum* (154).

NOTES (XI, 156–68)

156–8 Exclamations of grief, regret, and yearning fill these lines. Evander thinks of Pallas' youth (*primitiae iuvenis*), how the war so near his home (*propinqui*) brought bitter lessons (*dura rudimenta*) and how his own vows have been of no avail. This is the sum total of his agony of woe.

158 sanctissima - An adjective often used of the dead.

coniunx - Evander thinks of his wife, who died long before, and thinks her happy in her death because she has escaped such sorrow. We know little about Evander's Italian wife, the mother of Pallas, except that she was a Samnite woman (VIII, 510, *matre Sabella*). Some scholars say 'Sabine'.

160 vici...fata - 'I have outlived my allotted span of life'. Evander feels that he has lived too long.

160–1 The irony of Evander's long life is that he now survives his own son.

ut - Final use.

genitor - In an emphatic position at the end of the sentence.

161–3 Evander wishes that he could have gone to the war and lost his life in Pallas' stead, and thinks what the scene would have been like.

161 secutum agrees with *me* understood.

162 obruerent - The subjunctive may be interpreted either (1) as potential, 'the Rutuli would be overwhelming me...', or (2) as expressing a wish relating to past time, comparable to the normal construction with *utinam*, 'would that the Rutuli were overwhelming me...'. The imperfect tense makes Evander's thought all the more vivid: he seems to experience the fight in his desire to have suffered.

162–3 dedissem, referret are to be explained in the same way as *obruerent*: *dedissem* is in the pluperfect because it refers to the past, 'I should have...' or 'O that I had...', and *referret* is in the imperfect relating to the present, 'It would be bringing...' or 'O that it were bringing (now)...'.

pompa - The regular term for a funeral procession.

164 nec vos arguerim - The perfect subjunctive of a polite assertion: 'I would not think of...'.

165 ista implies a gesture towards the dead Pallas. It is because Evander has lived so long that he suffers the torment of seeing his son lying dead: he does not blame the Trojan alliance.

166–8 The thought that Pallas died full of honour supports him.

NOTES (XI, 167–76)

167–8 'First slaying thousands of Volscians, it will comfort (me) that he fell leading the Trojans into Latium.'

167 milibus – Poetic exaggeration for 'many'.

168 ducentem refers to *natum* in the preceding line.

cecidisse – From *cado*: accusative and infinitive construction with *iuvabit*.

iuvabit – Used impersonally. The thought will comfort Evander for the future.

169 quin introduces an intensified statement. Evander will rejoice in his son's glory, but even more in the honours paid to his dead body.

digner – Potential subjunctive, 'I would not honour you... (even if I could choose...)'.

funere – 'funeral honours'.

170–1 The repetitions of words and names echo Evander's pride which seems to mount as he speaks: notice the full-sounding *Phryges...Tyrrheni...Tyrrhenum*.

171 Tyrrhenum – Genitive plural.

172 'They bring great trophies, those whom your right hand consigns to death.' Evander sees displayed the arms which have been carried in procession (83) on tree-trunks according to the Roman custom. Supply *ei* as the antecedent to *quos*.

dat – The use of the present tense to indicate an action which has been done and remains as a fact. Pallas has slain them and remains their slayer.

173–5 Graphically addressed to Turnus. Evander utters his conviction that if Pallas and Turnus had been equally matched, Turnus would have been displayed as a tree trunk decked with his arms.

173–4 stares, esset – Subjunctives in a regular conditional sentence referring to present time and impossible of fulfilment.

174 si comes in meaning at the beginning of the line.

robur ab annis – It is unusual in Latin to find an adverbial phrase linked to a noun: 'strength from (his) years'.

175 Evander's thought is 'Why do I allow my selfish grief to detain the Trojans when they should be at the war?' He rallies from his passion of grief enough to think of the immediate future.

176 Addressed to the Trojans: they are to take messages back to Aeneas (*regi*).

memores agrees with *vos* understood from the imperatives.

NOTES (XI, 177–XII, 935)

177–9 'That I prolong the life I hate now since Pallas has been slain, your right hand is the cause, which you see owes Turnus to both the son and the father.' Evander is prepared to live longer only if he may see Aeneas avenge Pallas by slaying Turnus. These words are a dramatic prophecy of what is to come.

178 Turnum - Object of *debere*.

179 quam refers to *dextera*, and is subject of *debere*.

179–80 'This chance alone is left open for you, (to match) your deserts and your fortune.' Evander's meaning is that there is nothing else which Aeneas could more honourably and with more distinction accomplish than vengeance on Turnus.

180 vitae - Either genitive, 'joys of life', or dative, 'joys for life'; the latter is perhaps more poignant.

181 nec fas - Supply *est*. It would not be right for Evander ever to have joy after the loss of Pallas.

180–1 quaero governs two different constructions: (1) the accusative (*gaudia*); (2) an infinitive, *perferre*. Evander longs to carry the news of vengeance paid to Pallas in the underworld (*manes...imos*).

Aeneid XII, 930–52

930–8 These lines describe Turnus' unconditional and abject surrender to Aeneas. This is what Evander has longed and waited for (XI, 175–81).

930 Turnus is completely overwhelmed, as the words *humilis*, *supplex*, *precantem* serve to emphasize.

precantem - In an attitude of imploring for mercy.

932 utere - Imperative: governs the ablative *sorte*.

sorte - 'your fortune', i.e. Aeneas' right to slay Turnus.

932–3 Take in this order: *si qua cura miseri parentis te tangere potest*.

934 Dauni - Turnus thinks of his own father and begs for his life for his sake.

miserere - Imperative.

935–6 Turnus begs Aeneas at least to restore his body to his own people for burial, if he will not spare his life. The Romans used to attach great importance to proper burial in a man's own country.

935 lumine - Ablative of separation: 'the light (of life)'.

115

NOTES (XII, 936–47)

936 **victum** - supply *me*. Turnus acknowledges his complete and utter defeat.

937 **Ausonii** - A name used in a general sense for the native Italian tribes who made up the hosts commanded by Turnus. They are witnesses to his act of surrender as they surround the place where the single combat has taken place.

Lavinia - The cause of the war. The daughter of Latinus, king of the Latins, she was first betrothed to Turnus, but at the time when the Trojans were nearing Latium, Latinus was warned by an oracle that she was destined to marry a stranger. Latinus recognized Aeneas as Lavinia's future husband (*Aen.* VII, 81–106), but Turnus was not willing to relinquish her. His intractability was the cause of the struggle which subsequently broke out between Trojans and Latins.

938 'Further (than this) do not go in your hatred.' Turnus is ready to give up his claim to Lavinia and hopes, but vainly, that in return for this Aeneas will have mercy on him.

940 **iam iamque** - Used often in Vergil to describe a gradual process working up to a climax.

cunctantem refers to Aeneas.

sermo - Turnus' words.

941 **infelix** - 'ill-fated', 'unlucky' (x, 495–500). The belt catches Aeneas' eyes as he stands wondering whether to yield to Turnus' entreaties.

942 **balteus** - The sword-belt with the strap which fitted over the shoulder to hold the belt in place.

943 **victum** seems to recall *victum* referring to Turnus in 936. Turnus did not spare Pallas when he overwhelmed him, but yet begs Aeneas for his life when he finds himself in the same plight.

944 **straverat** - From *sterno*.

inimicum - 'belonging to his enemy'; there may also be an added meaning that the belt was hostile, i.e. would bring ill-luck to its wearer.

945 **monimenta** - The belt.

946 **exuvias** - Another expression for the belt. This word is used for anything stripped off the body.

947–8 'Are you, clad in the spoils of one I have loved, to be snatched away from me?'

947 **indute** - Vocative of the perfect participle: attracted to the case of *tu*, and giving a more pointed meaning.

NOTES (XII, 947–52)

meorum - Pallas: poetic plural.

948 **eripiare** = *eripiaris*: deliberative subjunctive, used when a person wonders what to do, or what is to happen.

949 Aeneas' words recall those of Evander (XI, 179–80) when he committed to Aeneas the act of vengeance which he lived to see accomplished.

immolat - 'sacrifices', a ritual word used for offering a victim to the gods. The slaying of Turnus is an atonement for the slaying of Pallas.

951 **ast** - Alternative form of *at*.

illi - Possessive dative.

frigore - The chill of death.

952 **indignata** - 'fretting' at its fate, that of death in early manhood; resenting the loss of Lavinia and the abject defeat before the assembled hosts. The *Aeneid* thus ends abruptly: Aeneas' work is done: Latium is secure for the Trojans who are the forbears of the Rome and Romans of the future.

NOTES ON BACKGROUND

Aeneid VIII

TIBERINUS

31–65. The Romans believed that all streams and rivers were sacred, and that gods or spirits lived in the waters. When Aeneas sleeps within the camp on the river bank, the god of the Tiber, Tiberinus himself, appears to him. This is fitting both because the Tiber is the chief river of Italy, and because it flows past the site of Rome. Acting as a link between Aeneas' camp and Pallanteum, Evander's settlement on the Palatine Hill, Tiberinus is gifted to speak prophetically of the future with the promise of eventual safety and a sure home for Aeneas and his people. Tiberinus, in appearance, is the very embodiment of his own river. He rises from among the poplars which fringe the river, trees which are commonly to be seen on river banks, especially in Italy, even today; his age calls for the reverence which must also be shown to the Tiber's stream; his white hair recalls the foam on water where it flows swiftly; he is clothed in a grey-green cloak, the colour of river-water. The linen of his cloak is of especial interest, as the word *carbasus* shows, which was used for a particular kind of flax which grew best in river-water. The thinness of his cloak suggests the flowing lines and transparency of water. He wears around his head a wreath of reeds from the river. The personification of rivers is a peculiarly Hellenistic conception adopted by the Romans and commonly used in both art and literature. As described by Vergil, Tiberinus has close parallels in several sculptured reliefs, which, although later than the Augustan age, are so similar as to allow the conjecture that Vergil envisages a traditional type. This thought was transmitted to English literature, as is instanced in Milton's lines (*Lycidas*, 103–4):

> Next Camus, reverend sire, went footing slow,
> His mantle hairy, and his bonnet sedge.

Here in the form of an old man wearing a cloak and a wreath of sedge is brought to life the river Cam, but he is almost the exact double of Tiberinus.

NOTES ON BACKGROUND

Troiana Urbs (Troia Nova)

36–7, 470–1. Part of Aeneas' mission in the west was to rebuild Troy in Italy. His ancestor Dardanus (134), the founder of the Trojan race from whom they received the names *Dardani* and *Dardanidae*, and after whom Troy was sometimes called *Dardania* (120), was believed to have come originally from Italy to Asia Minor. In this relationship Aeneas could be thought of as a native returned to his own land when he arrived in Italy, bringing Troy home again (37, *revehis*). As soon as Aeneas landed on Italian soil he built a camp for the safety of his followers on the left bank of the Tiber a little way in from the mouth. Vergil describes it in various passages as a strongly defended military camp with walls, towers and gates. This was the new Troy in Italy. It was situated exactly where Ostia, the port of Rome, was later to arise. In the ruins of the port which flourished until the end of the empire have been found traces of a small fort built of tufa blocks (the earliest building material used by the Romans) dating from the fourth century B.C. This *castrum* represents the earliest beginnings of Ostia. Vergil may have known of its existence and have thought of it as being built by Trojan invaders.

Alba

43–9. The prophecy of the sow was not new to Aeneas when he received it from Tiberinus, for he had heard it before in the same words during his wanderings in the eastern Mediterranean when he came to the town of Buthrotum on the west coast of Greece. There Helenus, the king, gave the Trojans hospitality, and a sacrifice was made on their behalf at the temple of Phoebus. The priest then pronounced an oracle (III, 374 ff.) which told that the sign of the settled home which Aeneas was seeking would be a white sow with a new-born litter of thirty young. It is fitting that Tiberinus should repeat this prophecy when Aeneas has at last reached the promised land, and only a little while before it is fulfilled. The sow with her miraculous litter was symbolic of the future, and closely connected in thought with the ancient towns of Lavinium and Alba. In the first place Aeneas was to found a city where the Penates, the gods he had brought from Troy, could be established (39). This was Lavinium, named after Aeneas' Italian bride, Lavinia, situated on the coast of Latium some twelve miles

NOTES ON BACKGROUND

south of Ostia. Ascanius would succeed him as king and after reigning there for thirty years, represented by the thirty young, would found another city called, after the white sow, Alba or, sometimes, Alba Longa. This was in reality one of the most ancient cities of Latium and, in legend, the immediate mother-city of Rome. It was situated among the Alban hills (the *colles Albani*) about fifteen miles to the south-east of Rome on a ridge overlooking the Alban Lake (*lacus Albanus*) and under the Alban Mount (*mons Albanus*), most probably near, or on, the site now occupied by the Pope's summer residence at Castelgandolfo. Although the city was destroyed in early times and the exact location long forgotten, its reality is undoubted. Vast cemeteries belonging to the early Iron Age of the eighth century B.C. have been excavated in this neighbourhood which prove the existence of an early settlement, and confirm the tradition of great antiquity. The legends told how, after the founding by Ascanius with a band of followers from Lavinium, kings ruled in Alba for three hundred years, until Romulus and Remus went from Alba to found Rome on the river-site by the Tiber. In this way Rome could be regarded as an Alban colony. Grown to power, Rome challenged the might of Alba and towards the end of the seventh century she was destroyed by the Roman king Tullus Hostilius. Thus the succession, according to the legends accepted in Vergil's time, was from Lavinium, through Alba, to Rome—from the coast, that is, to the hills and finally to the Tiber—and in this way the Romans were pleased to claim that they drew their origin from Trojan refugees who sailed to the coast of Latium after the sack of Troy. The name Alba is not in reality connected with *albus* ('white') as Vergil would have us believe, but is probably derived from the root *alb* ('mountain'): the usage here is an example of several fanciful derivations of place-names in the *Aeneid* which Vergil gives by means of false analogy but with a meaning fitted to the context.

NOTES ON BACKGROUND

Pallanteum and the Palatine Hill
(Mons Palatinus)

53-4, 98-101, 341. The legends concerning the founding of Rome all agree that the first settlement was established on the Palatine Hill. This is the most important of the hills of Rome: it was the nucleus of the ancient city and from it grew the city of the seven hills. It is a rocky plateau with precipitous sides, about twenty-five acres in extent, situated near the left bank of the Tiber, quite near the water and the island in the river. Before Rome was built and its formation obscured, the hill was isolated from the lower surrounding land by streams which flowed along deep valleys on each side to the Tiber. These were drained in early times and one of them became the Roman Forum. The hill was joined to the hinterland only by one narrow neck of land on the north-east, which was the Velia of Roman times. The Palatine, in its essential features typical of early sites in Latium, and easily defended by its rocky cliffs and difficulty of access, was chosen by the earliest settlers as the site of Rome as being a safe dwelling-place. Excavations penetrating down to the rock through layers of later ages have revealed, in several places, floors of beaten earth, post-holes and gutters together with black hand-made pottery (*impasto*) of a primitive type. These finds prove that the plateau was occupied in the first phase of the Early Iron Age by settlers who lived in simple huts. A particularly well defined hut floor recently excavated on the western edge is exactly comparable in form with Iron Age hut urns of terracotta (baked clay) found on many sites in Latium. These settlers, who are to be regarded as the primitive Romans, first occupied this hill, the very heart of Rome, in the eighth century B.C. From that time it was continuously inhabited through the successive ages until the end of the empire. During the centuries of the republic many temples were built on the Palatine, and from the fourth century B.C. onwards it began to be used for private houses. In the last century B.C. among several owners were Cicero, Crassus and Marcus Antonius. Then Augustus himself took over a house from Hortensius to be his own place of residence in Rome. This was the state of the Palatine Hill as it was known to Vergil, the first hill of Rome covered with temples and rich men's estates, but endowed with a legendary past belonging to the first beginnings of the city. In the *Aeneid* Vergil delights in thought to take away the present and to imagine the natural appearance of the land, painting a scene which, antedating the founding of Rome, goes back to the primeval shape of the earth. Although he had no archaeological knowledge he yet gives a glimpse of an early age which contains some elements of reality. Evander lives in a simple dwelling which recalls the huts of early ages; the view from the hill is that of nature's valleys and grassy slopes where the cattle low on the very site of Rome (360–1).

The origin of the name is unknown, but Vergil, interested always in tracing derivations of names, incorporated in his poem the legend current in his time, and earlier, that the first comers to the hill were Greeks. By a false but attractive analogy, he suggests that they came from Arcadia, and named the hill after their own city, Pallanteum. It is worthy of mention that the English word 'palace' is connected with *Palatinus*. An ancient stairway in the rock face which fronts the Tiber, called the *scala Caci*, a few traces of which can still be seen, gave rise to the legend of Hercules and the fire-breathing cattle-stealer Cacus, which has a place in the *Aeneid* (185–275). The *Lupercal*, a cave revered as sacred from the earliest days of Rome, also existed in the same side of the hill (see 'The Site of Rome', p. 130). Both in reality, as is proved by archaeological remains, and in legend the Palatine Hill was the cradle of infant Rome. Now it is covered with the ruins of vast substructures of palaces built by emperors long after Vergil's time, but on the western edge still remain primitive water tanks, and even traces of an Iron Age encampment to remind us of the ancient and humble origin of the great city.

The Navigation of the Tiber

86–101. Navigation up to Rome from the mouth of the Tiber was always difficult owing to the swiftness and strength of the current, and to the frequent meanders in the river's course. Ancient writers, especially the geographer Strabo, tell how even sails were of little use for taking boats upstream, and how merchandise had to be hauled up in barges from the docks at Ostia, the ancient port of Rome. In 1950 the writer navigated the Tiber downstream with some American friends in the month of August, from a point in Rome near where the oldest bridge, the *Pons Sublicius*, once spanned the river, at the foot of the Aventine Hill (*mons Aventinus*) near the place where Aeneas came to land, as far downstream as Ponte Galera, which is some four miles from Ostia. This journey was undertaken to aid in research into the navigation of the Tiber in primitive times. Although the land is comparatively flat all the way, the current proved to be strong enough to carry downstream a rubber raft without any added propulsion: steering only was necessary. It was clear that any attempt to row upstream would entail great physical effort. Aeneas and his companions could certainly not have reached Rome in a night and a morning, and this in the hot month of August, without the amelioration of the flow brought about by Tiberinus. The stilling of the stream was a divine favour which showed the blessing of heaven on Aeneas' mission; it is not a mere poetic trimming.

The Cult of Hercules in Rome

102–305. The worship of Hercules, essentially Greek in origin and character, was one of the first foreign cults to be introduced into Rome. It is most probable that it came in with Greek merchants from the cities of Magna Graecia in the south of Italy where the worship was widely practised, though some authorities think that it came through the medium of the Etruscans from the country north and west of the Tiber. At the foot of the Palatine, in the low-lying ground which extends from the Capitoline to the Aventine Hill and borders on the Tiber, there existed from early times a cattle-market, the so-called *Forum Boarium* (the name is connected with *bos*, an ox), which served as a place of trade and barter of all kinds. Close beside at the river's edge were the earliest docks belonging to republican times in the shape of simple moorings for ships. Traces of these can still be seen in the bank under the Aventine. This open space outside the city walls, frequented by foreign traders, was well suited to become the sanctuary of a god not native to the Romans but accepted by them. Hercules, the much-travelled god, had an especial connection with commerce and the men who brought their wares from overseas or from the hinterland beyond the Tiber. Here where foreign traders met was situated, near the river bank and at the eastern end of the Palatine Hill, a sacred grove. Within it stood the monuments of the cult of Hercules, a great altar known as the *ara maxima*, and a round temple. These were dedicated to him with the cult-title of *Hercules Invictus* (Hercules the Unconquered). The sanctuary could be seen from the river by anyone who arrived by water, just as Vergil describes Aeneas' arrival at the time of the sacrifice.

NOTES ON BACKGROUND

A realistic scene of the ritual of the yearly sacrifice to Hercules can be recognized in Vergil's description of the great festival which took place on 12 August. It shows characteristics of a Greek hero-cult. The *ara maxima* was the seat of a gentile cult (that is, privately owned and administered) belonging to two priestly families, the Potitii and the Pinarii. In this respect Vergil gives us the primitive picture, adding an archaic flavour to his account in the mention of the priest Potitius (281); by his own day the cult had become a state religion and was no longer the property of private families: history relates that the censor Appius Claudius appropriated it in 312 B.C. and handed the administration over to public slaves. From this time the *ara maxima* became the principal centre of the Hercules-cult in Rome. As Vergil tells us, it was even thought that Hercules himself had visited the spot and inaugurated his own worship. In the popular legend, told at length in the *Aeneid* (VIII, 185–275), Hercules reached the Palatine Hill driving the cattle of Geryones from Spain. A firebreathing monster Cacus, who inhabited a cave in the hillside, stole four of the oxen. Hercules recovered them and at the same time slew Cacus, who had been a danger to all the countryside. No doubt this tale was told on the occasion of the festival as Evander tells it to Aeneas: it is an aetiological myth composed to explain the existence of the Hercules-cult in that place. On the festal day a bull or a heifer was sacrificed in the morning and the praetor made libation with a wooden cup (*scyphus*) which recalls that often shown in Hercules' hand in Greek art. In the *Aeneid* this sacred act is carried out by Evander who acts as the praetor did in later times (VIII, 278). The sacrifice was followed on the next day by a banquet during which the worshippers were seated after the Greek manner: in a purely Roman festival they would have reclined for the feast. Women and slaves were excluded. In the evening the entrails of the victim were offered on the altar and there was a torch-light procession. The festival ended with another banquet in which the remains of the victim were entirely consumed. While the feasting continued, hymns were sung in honour of the hero-god, telling of his great deeds. Vergil describes faithfully practices which were to be seen and experienced yearly in his own time: we see with him all the ritual acts pass before our eyes, but to fit the times of which he wrote, before Rome was, he gives an archaic flavour in some of the details.

As the banquet proceeded far into the night, the *Salii*, the

NOTES ON BACKGROUND

dancing priests of Mars, chanted hymns in honour of Hercules, the hero-god. They told first the stories of his wonderful childhood, then how he sacked cities, and lastly recounted many of his great exploits (*labores*) done at the bidding of King Eurystheus. Hercules was the best known and most popular of Greek heroes, and because of his birth was worshipped as semi-divine. It was said that Jupiter visited Alcmena, wife of Amphitryon, disguised as her husband; Juno, the wife of Jupiter, jealous of the child who was subsequently born, sent two serpents to slay him, but the infant Hercules, already gifted with miraculous strength, strangled them both in his hands. Hercules sacked the city of Troy in his manhood, when after obtaining the girdle of Hippolyta he visited the city and found the inhabitants in distress. The walls of Troy had been built by the king Laomedon with the help of the gods Apollo and Poseidon, but now the king refused to give them the reward he had promised. The country was being ravaged by a sea-monster which could only be appeased by the sacrifice of Laomedon's daughter Hesione. Hercules offered to kill it in return for Laomedon's magical horses born of heavenly seed. Hercules carried out his part of the agreement, but Laomedon again dealt treacherously by refusing to give up the horses. In consequence, taking Hesione to be his bride, Hercules sacked the city. He destroyed the city of Oechalia when the king Eurytus refused to fulfil his promise to give him his daughter in marriage.

As a hero of great physical strength who travelled far and wide freeing mankind from the power of dangerous monsters, many myths were attributed to him in which he was said to have performed great feats of strength. Eventually these became crystallized in literature into a cycle of twelve so-called Labours which he was said to have carried out when, owing to Juno's continuing hatred, he had become for a time the slave of King Eurystheus. In the course of these exploits he slew the lion of Nemea and the hydra of Lerna, a water-snake with several heads; when one of these was severed two grew in its place. He hunted and caught the stag of Ceryneia, which had feet of bronze and antlers of gold, and the boar of Erymanthus which he presented alive to Eurystheus. He next shot with arrows the man-eating birds of the Stymphalian marsh and cleansed the Augean stables by diverting a river. For the seventh Labour he caught the bull of Crete which Poseidon had made mad because

NOTES ON BACKGROUND

Minos, the king, refused to offer it in sacrifice. He tamed the man-eating horses of Diomede, obtained the girdle of Hippolyta queen of the Amazons, and caught the three-bodied monster, Geryones. Then as his last two Labours he procured the golden apples from the garden of the Hesperides in north Africa, and dragged from the entrance to the underworld of the dead the guardian, a three-headed dog, Cerberus.

From these feats and many others attributed to him, Hercules is recognized as a type of hero who went about liberating men from dangerous beasts, a benefactor to mankind. In representations of him, including both sculpture and vase-painting, he is always shown with a club and a lion-skin, his own particular attributes. Although only four of the twelve Labours are mentioned by name in the hymn of the *Salii* (293–302), the Cretan bull, the Nemean lion, Cerberus, and the hydra of Lerna, it is implied that many others were told. The incident of the slaying of Hylaeus and Pholus is an example of the many *parerga*, minor exploits, which were told about him. This was attached to the hunting of the Erymanthian boar. During his journey, Hercules stayed to rest with the centaur Pholus, who opened for his benefit a bottle of good wine. Attracted by the smell of the wine, several other centaurs set upon them, and in the ensuing skirmish Hercules slew several including Hylaeus and, by great misfortune, his host Pholus. The slaying of the giant Typhoeus is another *parergon*. By selecting for mention in the hymn certain of the twelve Labours and certain minor deeds, Vergil hints at the great cycle of myth and legend which centred round the figure of the semi-divine hero.

THE SALII

285–305. The feasting with which the sacrifice to Hercules culminated at the end of the day was enlivened by the appearance of the *Salii* who, in choral songs and dances, celebrated the great deeds of the hero-god. Their inclusion in the rites of Hercules shows that Vergil was aware of the great antiquity of the institution, and even attributed it to an age older than Rome. The name was given, in Roman times, to one of the oldest and most important priestly brotherhoods in the state religion. This was divided into two colleges, one of which was closely connected with the Palatine Hill (*Salii Palatini*). It seems fitting that Vergil should bring these within the ritual of the *ara maxima* since it lay

126

NOTES ON BACKGROUND

close under the hill. The name, connected with *salire* ('leap'), indicates their religious duties. At fixed times during the year they performed sacred dances of a warlike and primitive character, in the full military dress of a warrior of earlier times. They wore an embroidered tunic with a wide girdle of bronze, and a *trabea*, a short cloak striped with scarlet and bordered with purple, and a heavy bronze corselet: at their side they carried a sword, and wore a cone-shaped helmet (*apex*) of an ancient type. The ceremonies in which as war-priests they were required to take part fell in March and October, at the beginning and end of the Roman campaigning season, and their especial festivals were the *Quinquatrus* (19 March) and the *Armilustrium* (19 October). At the beginning of March they would carry in procession round the city for several days the *ancilia* from the shrine of Mars in the Forum, shields of an antique figure-of-eight shape, said to be copies of the original *ancile* sent down from heaven by Jupiter as a gift to Numa, the second king of Rome. As they went along, they stopped at prescribed places to perform their strange ritual dances, leaping, beating the shields and chanting a litany, the *carmen saliare*, so ancient that the meaning was lost to the Romans themselves. A few fragments still exist but are in such old Latin as to have become meaningless. In October at the ceremony of the *Armilustrium* the shields were replaced. Twenty-four men composed the brotherhood, which was dedicated to the war-god Mars: membership was a privilege granted only to young men of patrician birth and was considered so exclusive that no other religious office might be held at the same time.

The Site of Rome

310–61. Evander takes Aeneas to his home on the Palatine Hill (*mons Palatinus*) from the place where the sacrifice to Hercules has been observed. This was the open low-lying ground which bordered on the Tiber and was half encircled by three hills, the Aventine (*mons Aventinus*), the Palatine, and the Capitoline (*mons Capitolinus*): it was to become the cattle-market of Rome and was to be known as the *Forum Boarium* (connected with *bos*, 'ox'). Evander talks, as they go, first about the bygone ages of Latium (314–36), then points out to Aeneas on the site of future Rome, as they look at rocks, grassy slopes and valleys, climb tree-clad hills and see ruins left by men of times more ancient even than those of Evander, places revered for their sanctity through all the days of Rome.

First they walk along the outside of the wall, which runs along the foot of the three hills. Rome was surrounded with a strong defensive wall after the Gallic invasion in the fourth century B.C.: traces of this still survive in several quarters of Rome, and it may well be that in Vergil's day it remained intact and was thought to be very ancient. There can be no doubt, however, from the fact that Evander and Aeneas come first to a gate, that Vergil thought of the *Forum Boarium* as outside the wall. They come to the gate and altar of Evander's mother, Carmentis (338), which provided one of the entrances through this wall into the city. It was much used in Roman times by those who came up the Tiber, and landed, just as Aeneas did, at the docks on the river's edge. The altar and the gate stood side by side. Carmentis was a native Italian goddess of prophecy, as her name, connected with *carmen* ('song') because prophecies were usually chanted, clearly shows. Vergil brings her into Evander's story by identifying her with his mother, now transformed into a nymph, who caused him to slay his father. Remains of this altar of a primitive type still exist beside a stretch of paved road which represents the path by which Evander brings Aeneas to his first view of the natural, primeval landscape where future Rome is to be built. Once through the gate, they look at the nearer view where stand the two most important and majestic of the seven hills of Rome. On their left are the steep sides of the Capitoline and on their right the Palatine where Evander has his home. The Capitoline stands, a precipitous height, with sheer rocky cliff-like sides; and there are two eminences on the summit. There can be no doubt that the higher of the two on the north-eastern edge was the *arx* of primitive Rome, the citadel or 'defensive' place. Between them was the dip known as *inter duos lucos* ('between the two groves'), where, the legend went, Romulus the founder of Rome made a place of safety, the *asylum* (a name derived from the Greek meaning 'refuge' or 'sanctuary'), for refugees and outlaws from the surrounding country. These, it was said, were the first inhabitants of his city. Evander, knowing nothing of Romulus, only shows Aeneas a clearing in the forest (*lucus*) which clothes the hill. When Vergil mentions Romulus (342) he speaks direct to all Romans of his day to whom the name is sacred. After pointing upwards to the place of the *asylum* Evander points across to the right, to the

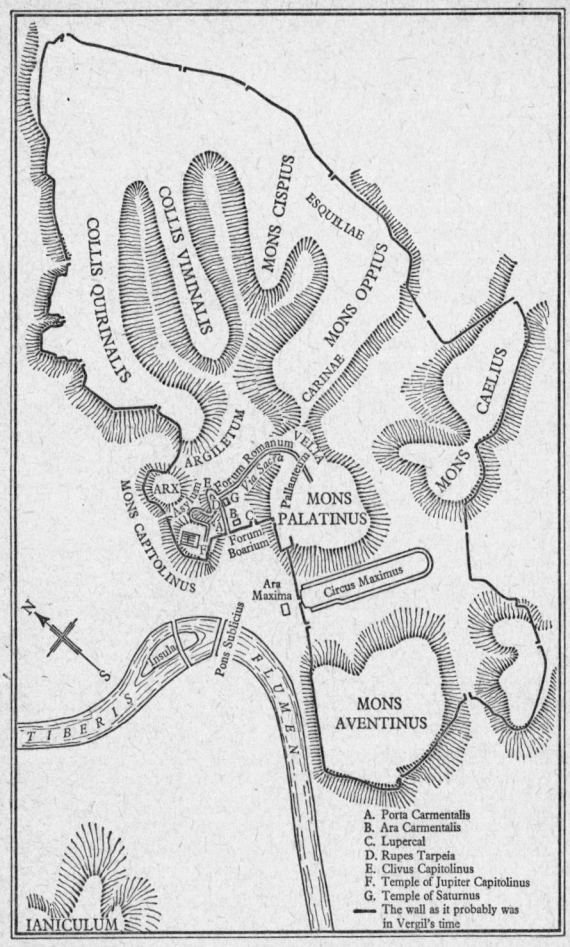

Plan of Ancient Rome to illustrate Vergil's *Aeneid* VIII

Palatine, and calls Aeneas' attention to a cave in the cliff-like face of the hill in the south-west corner. Then merely a natural grotto, it was to be one of the most sacred spots in all Rome, the *Lupercal* (connected with *lupus*, 'wolf'). It was in this cave, so the ancient story went, that the twins Romulus and Remus were nursed by a she-wolf. In the foundation-legend of Rome it was told how a priestess from Alba, by name Rhea, bore twin boys to the god Mars. Her uncle was Amulius, the usurping king of Alba who had seized the kingdom from her father Numitor. Amulius feared that when the boys grew up they might try to avenge their father and to restore him to his rightful throne. To do away with this possibility he exposed the babies on the bank of the Tiber. They were saved because the river was in flood and floated them gently into the cave. A shepherd, Faustulus, found them there suckled by a she-wolf, and took them to his hut and cared for them until they grew to manhood. So it was that Romulus became the founder of Rome. This cave was venerated through all the ages of the city from the earliest beginnings, and was beautified with monuments and a sculptured relief of the wolf and the twins. So sacred was the *Lupercal* in Roman thought that it was one of the many shrines which Augustus renewed in his endeavour to strengthen the old cults of Rome.

Evander next shows Aeneas the landscape on their left. They look at a wooded glade; it is the *Argiletum* (345), which was to become a crowded quarter of Rome crossed by a busy street. Remains of the paving-stones deeply rutted by the wheels of carts can still be seen on the edge of the Roman Forum. By a fanciful derivation of the name Vergil suggests that it means 'the death of Argus'. He had once been Evander's guest, but, when he plotted to kill the king and to take his place, he was put to death by Evander's loyal subjects. This was without the king's knowledge or consent, so it is no wonder that, looking at his place of burial, Evander protests his own innocence (346). The word is most probably connected with *argilla*, 'potter's clay' with the suffix *-etum*, and means no more than 'a claypit'.

Evander now leads Aeneas to the Capitoline Hill, and, as they pass under it, shows him the frowning precipitous rock at the south-eastern corner called the *rupes Tarpeia* ('the Tarpeian Rock'). Vergil calls it *Tarpeia sedes* (347), which was a place of execution. Ancient writers tell how criminals were dashed down to death below in full view of crowds watching in the Forum.

NOTES ON BACKGROUND

Legend told how the girl Tarpeia gave her name to the rock. When Romulus was warring with the Sabines, he entrusted the defence of the Capitoline to her father Tarpeius. His daughter went out to fetch water, but fell into the enemies' hands. Attracted by their gold bracelets, she offered to show them the way into the stronghold in return for what they had on their left arms. For this betrayal she did not receive the reward she had asked, but did receive what they carried on their left arms, for they slew her by battering her to death with their shields. They buried her on this spot and so she gave her name to the rock.

After gazing at this grim sight Evander and Aeneas climb the hill and reach the summit (347-58), where they survey the scene. They ascend by a steep, rough path which winds up to the top. A paved road now covers this path, the *clivus Capitolinus* of the Romans, one of the most ancient of all the ways in Rome. Parts of it have been uncovered in recent years to lay bare the very stones over which the Romans used to pass. They reach the eminence on the south-west of the Capitol where later was to be built the greatest of all the temples in Rome, that sacred to Jupiter Capitolinus. This temple was originally built in Etruscan times when the kings ruled in Rome. When the foundations were being dug it was said that a human head of gigantic size was found and that this was a sign that Rome would some day rule the world. It was rebuilt several times, but in Vergil's own day the roof was covered with plates of gilded bronze: he mentions its appearance in the word *aurea* (348). This temple was the centre of the religious life of the Roman state. Here consuls offered sacrifice on taking up office, triumphal processions climbed the *clivus Capitolinus* to halt at its doors, archives of state were deposited within, and every Roman boy of good family came here to make offering on reaching manhood. Remains of the high platform of huge tufa blocks can still be seen, testimony of the size and importance of this building.

The view extends to the country across the Tiber where the Janiculum hill confronts the spectator. Evander shows Aeneas remains of settlements belonging to an age even earlier than that of the Arcadians. They look at walls left by primitive men on both the Capitoline and the Janiculum. It is possible that in Vergil's own time ruined walls and houses were to be seen there which were thought to be older than Rome. There was in reality, however, in Roman times an important temple dedicated to Saturn on the lower slopes of the Capitol, facing the Forum: from this the hill itself was often called *Saturnius*. It was easy for a legend to grow up that there was once a town on the hill of the same name.

They descend the Capitoline by the same path, for there was no other means of access to the top, and make their way along the valley which extends from the foot in an easterly direction. They pass herds grazing peacefully in grassy pastures where was to be the Roman Forum, the vital heart of the future city. The path which they follow to reach Pallanteum is that which became the *sacra via* ('the Sacred Way'), which ran from the Capitoline through the whole length of the Forum and curved at the farther end to reach the top of the Palatine. As they reach the hill they look across to the *Mons Oppius* ('the Oppian') which is the southern ridge of the *Mons Esquilinus* ('the Esquiline'): here are grass-grown slopes given over to flocks and herds. In Vergil's time this was a fashionable residential quarter called *Carinae*: the name was said to have come from certain buildings which resembled keels of ships.

All that Evander shows to Aeneas belongs to the very heart of the oldest part of Rome, and to the earliest memorials of human settlement on the site. The hills, valleys, rocks and pastures are in the future to be built over with the paved streets, halls, temples, markets and palaces which made Rome one of the wonders of the ancient world. Vergil paints a pre-Roman landscape, in which are the natural features which will become the most sacred places in the Roman world, symbols of the might and majesty of the dominion of Rome, and yet inherent in her first beginnings.

AGYLLA (CAERE)

478–80, 597–601. The origin of the Etruscans (480) still remains a question unsolved alike by historians and archaeologists, for some hold, with the ancient authorities (among them Vergil himself, 479), that they came from Lydia on the coast of Asia Minor, others that they were Italic folk moulded by influences from the East which eventually gave them their distinctive national character. During the time of their greatest power they were in possession of the land which extends between the Tiber, the Arno, and the sea, and which in consequence was often called Etruria. It was said, moreover, that nearly the whole of Italy was once under Etruscan domination, and it is certain that

NOTES ON BACKGROUND

Etruscan kings ruled in Rome until the establishment of the republic. From the eighth until the sixth century B.C. they were the lords of Italy. Twelve powerful cities made up the Etruscan league, of which one, by no means the least in strength, was the city called in Greek Agylla (479), in Etruscan Caere (597).

Caere, as it is more usually called, was situated on a ridge of tufa rock (the natural volcanic rock of all this part of Italy) about twenty miles to the north of Rome, and about five inland from the Tyrrhenian sea, and can still be visited at the present day. The traveller approaches the site by the Aurelia, the old Roman road which is now the modern highway out of Rome. At a small distance away from the village he crosses a small stream called the Vaccina, which flows down a pleasantly wooded valley backed by slopes covered with dark brushwood; this is Vergil's *Caeritis amnem* (597). Somewhere near by was the grove of Silvanus where Aeneas with Pallas joined Tarcho and his followers encamped outside the city and rested after his twenty-mile ride from Pallanteum (600-7). A little further on the entrance of the modern village is reached, the name of which, Cerveteri, from *Caere Vetus*, preserves the ancient Etruscan. Passing through a medieval gateway, the traveller sees a squalid little place which occupies only a fraction of the extent of the once splendid city, which in its prime extended over a plateau of rock some four miles in circumference. It was then surrounded by a wall, of which some fragments remain, and had at least eight richly decorated temples within its *enceinte*.

The elevated position gave the city a commanding position over the neighbouring coast where three harbours at least were hers. In every direction extend the fertile plains still given over to crops and pasture, from which she drew her livelihood. Cerveteri is famous, and much visited today not so much for the archaeology of the city as for the vast cemeteries with which the site is surrounded. Remains from tombs revealed by excavation, ranging in date from the Iron Age of the eighth century B.C. (the so-called 'Villanovan' phase) onwards, testify to abundant wealth, and to trade overseas with Greece and the East. This prosperity reached its peak in the seventh and sixth centuries B.C. Beehive tombs cut out of the living rock and sculptured within in the shape of houses, some of which are even furnished with household objects, have much to tell of the architecture and the daily life of the Etruscans of Caere. They are unique in all

NOTES ON BACKGROUND

ancient Etruria. It is possible, too, to walk down a ravine which is none other than a street of tombs which are cut out of the living rock on either side, to climb over the entrances and see still left on the funeral benches sherds of pottery, perhaps of the finest that fifth-century Athens produced. Pottery found in great quantity in many of the tombs shows unbroken trade with Greek cities through several centuries: some may be seen in the British Museum, as fine as any that ever came out of Greece. The Regolini-Galassi tomb, so called after the two Italians who excavated it, is of great size and belonged to some rich Etruscan overlord. It yielded exquisite ornaments of pure gold for the body, masterpieces, in design and execution, of the goldsmith's art, finely worked arms, cauldrons and vessels of bronze, and quantities of carved ivories. Air photography has recently revealed vast numbers of tombs yet to be explored: it has been calculated that more than two thousand exist which cannot be seen at ground level. In Vergil's own day Caere was a small Roman town: her Etruscan character and her greatness, long departed, remained only in tradition and in literature.

Praeneste

561. Praeneste, modern Palestrina, was one of the most powerful cities in Latium, second only to Rome, during the time of her *floruit* which lasted from the seventh century B.C. to the third century A.D. The city was set on the slopes of Rocca Ginestra, a spur of the Montes Praenestini, in a position of great natural strength which guarded the pass from the valley of the Sacco to the lower reaches of the Tiber. Standing only some twenty-three miles from Rome, the most ancient part of the city which archaeology shows as dating from the early Iron Age had for a citadel the summit of the hill. Ancient walls of polygonal masonry, but belonging to a later age, join this to the lower part of the city at the foot of the hill.

Evander's struggle with the three-bodied monster Erulus (563) seems to foreshadow the course of history. Until the last century B.C., although subdued on several occasions by the Romans, Praeneste was able to maintain a certain independence. The natural strength of her position gave her an advantage not enjoyed by other peoples who fell before the advancing power of Rome in the fourth, third and second centuries B.C. In the last

VOCABULARY

Diphthongs, final -*i*'s and -*o*'s are long (with a few exceptions which are indicated); these long vowels are not marked. All other long vowels are marked, and any unmarked vowel can be assumed to be short. For long *syllables* in verse see pages xxii and xxiii in the Introduction.

ABBREVIATIONS

1, 2, 3, or 4 after a verb means that it is a regular verb like *amo*, *moneo*, *rego* (*capio*), or *audio*, unless the perfect and supine are given.

abl.	ablative	interj.	interjection
acc.	accusative	interrog.	interrogative
adj.	adjective	intr.	intransitive verb
adv.	adverb	m.	masculine
c.	common	n.	neuter
comp.	comparative	part.	participle
conj.	conjunction	pass.	passive
dat.	dative	pl.	plural
defect.	defective verb	p.p.p.	past participle passive
dep.	deponent verb	prep.	preposition
f.	feminine	pron.	pronoun
gen.	genitive	subj.	subjunctive
impers.	impersonal verb	superl.	superlative
indecl.	indeclinable	tr.	transitive verb
indef.	indefinite		

NOTES ON BACKGROUND

century B.C. Praeneste was the scene of Sulla's victory Marius: from then onwards she enjoyed great fame and sperity. In an earlier age, however, during the time Etruscan domination, Praeneste had also been outstand prosperous under the rule of wealthy overlords.

The contents of tombs in the great cemetery at the foot of hill make this abundantly clear. Especially famous is treasure from a great tomb of the seventh century B.C. (Barberini tomb) which for magnificence and splendour can o be compared with those found at Caere (see p. 134). Exquisit worked ornaments of purest gold for the body of the dead, togeth with vessels of silver and bronze and quantities of carved ivor show a flowering beyond count. This time of riches and wealth wa only a half-remembered tradition in Vergil's day. One monument however, to which practically the whole of the hillside was dedicated, made Praeneste famous not only in Latium, but throughout the Roman world. This was the great Temple of Fortune which by Vergil's time had become a fashionable and much frequented sanctuary. The origin of the cult goes back to great antiquity: it can be recognized in an oracular cave still to be seen in the pillared hall at the foot of the hill. There *sortes* were given by the priests on ancient lettered tablets which were said to have been found in the rock. The whole temple site was enriched with architecture in the second century and again by Sulla as a thank-offering to the goddess *Fortuna* for his victories. Much of this has always been known, although a medieval and modern village was built into the ruins.

During the last war Palestrina was heavily bombarded. The effect was to strip the hillside of all but the Roman structures: the precincts of the cult-site mounting up the hillside from one level to another from foot to summit here appeared, set against a rock face covered with polygonal or *incertum* masonry. A series of ramps, colonnades and staircases, platforms and arched rooms mounting up to a pillared hemicycle which crowned the whole scheme were uncovered. This monumental work, now greatly restored, is a masterpiece of Hellenistic architecture and shows the planning of some genius of the ancient world. This was the Praeneste known to Vergil.

VOCABULARY

ā, ab, prep. with abl. *from*

Abās, -ntis, m. *an Etruscan warrior slain by Lausus*

abiēs, -etis, f. *pine, fir, pine- or firwood, ship, spear*

abrumpo, 3, -rūpi, -ruptum, tr. *break off, cut short*

abscēssus, -ūs, m. *departure*

absisto, 3, -stiti, intr. *abstain from*; **absiste movēri**, *be not alarmed*

abstuli, see **aufero**

absum, -esse, āfui, ——, intr. *be away (from), be distant*; **absēns**, -ntis, *absent, in one's absence*

āc, conj. *and, as, than*

accēdo, 3, -cēssi, -cēssum, intr. *come, approach, be added*

accendo, 3, -di, -sum, tr. *kindle*

accipio, 3, -cēpi, -ceptum, tr. *receive, hear*

ācer, ācris, ācre, adj. *keen, brave, fierce*

acerbus, -a, -um, adj. *bitter*

acernus, -a, -um, adj. *of maple-wood*

acervus, -i, m. *heap, pile, mass*

Achātēs, -ae, m. *comrade of Aeneas*

aciēs, -ēi, f. *troop, edge, battle-line, rank*

Acoetēs, -ae, m. *a Trojan warrior*

acūtus, -a, -um, adj. *sharp*

ad, prep. with acc. *to, towards*

addēnseo, 2, tr. *thicken, close up, make close*

addo, 3, -didi, -ditum, tr. *add*

addūco, 3, -dūxi, -ductum, tr. *draw up, lead on*

adeŏ, adv. *so far, so much, indeed*

adeo, -īre, -ii, -itum, tr. and intr. *approach*

adfātur, 1, dep. defect. tr. *speak to, address, accost*

adhibeo, 2, tr. *bring towards*

adhūc, adv. *as yet, until now*

adicio, 3, -iēci, -iectum, tr. *add to, tell in addition*

adimo, 3, -ēmi, -emptum, tr. *take away, deprive of*

adiungo, 3, -iūnxi, -iūnctum, tr. *join to, add to*

adiuvo, 1, -iūvi, -iūtum, tr. *assist, help, aid*

adlābor, 3, -lāpsus, dep. intr. *glide towards* (with dat. or acc.)

adloquor, 3, -locūtus, dep. tr. *address, speak to, accost*

adluo, 3, -ui, tr. *wash against*

adsiduē, adv. *constantly, continually, without ceasing*

adsisto, 3, -stiti, intr. *stand beside*

adsuēsco, 3, -suēvi, -suētum, intr. *become used to*, tr. *accustom*

adsum, -esse, -fui, ——, intr. *be present, aid, help*

adveho, 3, -vēxi, -vectum, tr. *bring or carry to a place*; pass. *ride or sail to*

advena, -ae, m. *stranger*

adversus, -a, -um, adj. *opposite, facing, fronting*

adverto, 3, -ti, -versum, tr. *turn to or towards, heed*

advolo, 1, intr. *fly to, rush at*

aedēs, -ium, f. pl. *dwelling-place, house, home*

aegis, -idis, f. *shield (of Jupiter)*, acc. **aegida**
aemulus, -a, -um, adj. *rivalling*
Aeneadae, -um, m. pl. *followers of Aeneas*
Aeneadēs, -ae, m. *son or descendant of Aeneas*
Aenēās, -ae, m., *son of Venus and Anchises, leader of the Trojans; the chief character in the story of the Aeneid*
Aeolius, -a, -um, adj. *Aeolian*
aequē, adv. *equally*
aequo, 1, tr. *make level*
aequor, -oris, n. *plain, sea, level surface*
aequus, -a, -um, adj. *equal, fair, level*
aes, aeris, n. *bronze*
aestās, -ātis, f. *summer*
aetās, -ātis, f. *age, time*
aeternus, -a, -um, adj. *lasting*
aethēr, -eris, m. *air, sky, heaven* (acc. **aethera**)
aetherius, -a, -um, adj. *heavenly*
Aethōn, -ōnis, m. *Pallas' war-horse*
aevum, -i, n. *age, life*
āfore, see **absum**
ager, agri, m. *land, field*
aggero, 1, tr. *pile or heap up*
agmen, -inis, n. *troop, band*
agnōsco, 3, -nōvi, -nitum, tr. *recognize*
ago, 3, ēgi, āctum, tr. *drive, lead, do, accomplish, discuss*
agrestis, -is, m. *farmer, countryman, rustic*
Agyllīnus, -a, -um, adj. *of or belonging to Agylla*
āio, ait, defect. *say, assert, affirm*

āla, -ae, f. *wing, pinion*
Alba, -ae, f. *the mother-city of Rome*
Albula, -ae, f. *ancient name of the Tiber*
albus, -a, -um, adj. *white*
Alcīdēs, -ae, m. *Hercules, descendant of Alceus*
āles, ālitis, c. *bird;* adj. *winged, quick, swift;* gen. pl. **ālituum**
aliquandō, adv. *at some time, at last*
aliquis, -quid, indef. pronoun, *someone, anyone*
alius, -a, -ud, adj. *other;* pl. *some...others*
almus, -a, -um, adj. *gracious*
alo, 3, -ui, altum (alitum), tr. *feed, support, nourish, nurture*
altāria, -ium, n. pl. *altar*
altus, -a, -um, adj. *high, deep*
amārus, -a, -um, adj. *bitter*
ambiguus, -a, -um, adj. *doubtful, uncertain, unsettled*
ambō, -ae, -o, pl. adj. *both*
amictus, -ūs, m. *cloak, mantle*
amīcus, -a, -um, adj. *friendly*
āmitto, 3, -mīsi, -missum, tr. *lose*
amnis, -is, m. *stream, river*
amoenus, -a, -um, adj. *lovely, pleasant, charming* (usually describes scenery)
amor, -ōris, m. *desire, love*
Amphitryōniadēs, -ae, m. *son of Amphitryo, Hercules*
amplector, 3, -plexus, dep. tr. *clasp, embrace, enfold*
amplexus, -ūs, m. *embrace*
an, conj. *or, or whether*
Anchemolus, -ī, m. *Latin warrior*

VOCABULARY

Anchīsēs, -ae, m. *father of Aeneas*; **Anchīsiadēs**, -ae, m. *son of Anchises, Aeneas*
anguis, -is, m. *snake, serpent*
angustus, -a, -um, adj. *narrow*
animal, -ālis, n. *a living thing*
animus, -i, m. *mind, thought*
annus, -ī, m. *year*
annuus, -a, -um, adj. *yearly*
ante, adv. *before, formerly*; prep. with acc. *in front of*
antrum, -i, n. *cave, cavern*
aperio, 4, aperui, apertum, tr. *open*
apertus, -a, -um, part. as adj. *open, exposed*
Apollo, -inis, m. *Apollo, god of the sun, the bow, and of prophecy*
appāreo, 2, -ui, -itum, intr. *appear, be seen*
apparo, 1, tr. *prepare*
apto, 1, tr. *fit, man*
āra, -ae, f. *altar*
arbor, -oris, f. *tree, wood*
arbustum, -i, n. *bush, shrub*
arbuteus, -a, -um, adj. *of arbutus* (wild strawberry tree)
Arcadia, -ae, f. *Arcadia, district in Greece*
Arcas, -cadis, m. *an Arcadian*
arceo, 2, -ui, arctum, tr. *prevent, ward off, keep off*
Ardea, -ae, f. *capital of the Rutuli*
ārdeo, 2, ārsi, ārsum, intr. *be on fire, blaze, be eager*
arduus, -a, -um, adj. *high*
Argīlētum, -i, n. *a district in Rome adjoining the Roman Forum*
arguo, 3, -ui, -ūtum, tr. *upbraid, rebuke, accuse*
Argus, -i, m. *guest of Evander*
arma, -ōrum, n. pl. *arms*
armentum, -i, n. *herd, cattle*
armiger, -eri, m. *armour-bearer*
armo, 1, tr. *arm*
ars, artis, f. *art, skill*
ārsuras, see ārdeo
artūs, -uum, m. pl. *limbs*
arvum, -i, n. *field, land*
arx, arcis, f. *citadel, stronghold*
Ascănius, -ii, m. *son of Aeneas*
asper, -era, -erum, adj. *rough, rugged, fierce, rude*
ast or **at**, conj. *but*
astrum, -i, n. *star*
asȳlum, -i, n. *refuge, sanctuary*
at, conj. *but*
āter, ātra, ātrum, adj. *black*
Ătlantis, -idis, f. *daughter of Atlas, Electra*
Ătlās, -antis, m. *Atlas*
atque, conj. *and*; **haud secus atque**, *not otherwise than, just as*
Ătrīdēs, -ae, m. *descendant of Atreus*
attollo, 3, —, —, tr. *raise, lift up*; pass. *rise*
auctor, -ōris, m. *adviser, founder*
audāx, -ācis, adj. *daring, bold*
audeo, 2, ausus, semi-dep. intr. *dare, have courage to*
audio, 4, -īvi (-ii), -ītum, tr. *hear, heed, listen to*
aufero, -ferre, abstuli, ablātum, tr. *carry off, take away*
aura, -ae, f. *air, wind, breeze*
aureus, -a, -um, adj. *golden*

VOCABULARY

auris, -is, f. *ear*
aurum, -i, n. *gold*
Ausonius, -a, -um, adj. *Ausonian, Italian*
auspicium, -i, n. *omen, leadership*
ausus, see **audeo**
aut, conj. *or*; **aut...aut**, *either...or*
autem, conj. *but, moreover*
auxilium, -i, n. *help, succour*
axis, -is, m. *pole, zenith*

Bacchus, -i, m. *Bacchus, wine-god, wine*
balteus, -i, m. *belt*
bellātor, -ōris, m. *warrior, fighter*; **bellātor equus**, *war-horse, charger*
bellum, -i, n. *war, contest, strife*
bidēns, -entis, f. *having two teeth* (i.e. animal for sacrifice, usually sheep)
biiugi, -ōrum, m. pl. *two horses yoked, a pair, a two-horse chariot*
bimembris, -e, adj. *half-man, half-beast*, adjective applied to centaurs
bīni, -ae, -a, pl. adj. *two each, two, a pair of, double*
birēmis, -is, f. *boat with two banks of oars*
bōs, bovis, c. *ox, cow* (gen. pl. boum)
brevis, -e, adj. *short, brief*
bulla, -ae, f. *stud*

Cācus, -i, m. *giant, son of Vulcan, slain by Hercules*
cado, 3, cecidi, cāsum, intr *fall, set* (of stars)
caedēs, -is, f. *slaughter*
caedo, 3, cecīdi, caesum, tr. *strike, slay*
caelestis, -e, adj. *divine, on high*
caelo, 1, tr. *engrave*
caelum, -i, n. *sky, heaven*
Caere, -itis, f. *Etruscan city*
caeruleus, -a, -um, adj. *dark blue, dark-coloured, dark green*
caesus, see **caedo**
calidus, -a, -um, adj. *hot*
calx, calcis, f. (rarely m.) *heel*
campus, -i, m. *plain, field*
candidus, -a, -um, adj. *gleaming white, bright, fair*
canis, -is, c. *dog*
canistrum, -i, n. *basket*
cano, 3, cecini, cantum, tr. and intr. *sing, chant, prophesy*
cantus, -ūs, m. *song, charm, chant*
capesso, 3, -īvi (-ii), -ītum, tr. *try to reach, make for, seize*
capio, 3, cēpi, captum, tr. *take, seize, delude, mislead*
Capitōlium, -ii, n. *the Capitol*, one of the seven hills of Rome (sometimes pl.)
caput, -itis, n. *head, tip, end, chief, life, person*
carbasus, -i, f. *very fine Spanish flax, fine linen cloak*
carīna, -ae, f. *keel, ship*
Carīnae, -ārum, f. pl. *a quarter of Rome near the Esquiline Hill*
carmen, -inis, n. *song, chant*
Carmentālis, -e, adj. *belonging to Carmentis*
Carmentis, -is, f. *mother of Evander, a prophetess*
cārus, -a, -um, adj. *dear*

140

VOCABULARY

castra, -ōrum, n. pl. *camp*
cāsus, -ūs, m. *event, chance*
causa, -ae, f. *cause, reason*; in abl. with gen. *for the sake of*
cautus, -a, -um, adj. *careful*
cavus, -a, -um, adj. *hollow*
cecidēre, cecidisse, see **cado**
cēdo, 3, cessi, cessum, *give way*
celebro, 1, tr. *throng, attend, celebrate, frequent*
celero, 1, tr. and intr. *hasten*
cēlo, 1, tr. *hide, conceal, keep secret*
celsus, -a, -um, adj. *high, lofty*
centum, indecl. num. adj. *hundred*
cerebrum, -i, n. *brain*
Cerēs, -eris, f. *Ceres, goddess of agriculture, corn*
cerno, 3, crēvi, crētum, tr. *see, perceive*
certāmen, -inis, n. *quarrel, strife, struggle*
certātim, adv. *in rivalry, eagerly*
certus, -a, -um, adj. *sure*
cēterus, -a, -um, adj. *remaining*; pl. *the rest*; nom. sing. masc. not used
chlamys, -ydis, f. *cloak*
chorus, -i, m. *band of dancers*
cieo, 2, cīvi, citum, tr. *stir*
cingo, 3, cīnxi, cīnctum, tr. *clothe, gird, encircle*
cingulum, -i, n. *belt, girdle*
circum, adv. and prep. with acc. *around*
circumdo, 1, -dedi, -datum, tr. *surround, encompass, wrap around*
circumsisto, 3, -steti, tr. *stand around, beset*
circumsto, 1, -steti, tr. and intr. *throng round, stand around*
cīvis, -is, c. *citizen, fellow-countryman, compatriot*
clāmor, -ōris, m. *cry, clamour*
clangor, -ōris, m. *clang, bray or call* (of trumpeter)
clārus, -a, -um, adj. *bright, clear, famous, noble*
classis, -is, f. *fleet*
claudo, 3, -si, -sum, tr. *shut, hem in, surround*
clipeus, -i, m. *round shield*
Clonus, -i, m. *a goldsmith*
coeo, -īre, -īvi (-ii), -itum, irreg. intr. *join, come together*
coepi, -īsse, -tum, defect. tr. *began*
coeptum, -i, n. *plan, enterprise, attempt, undertaking*
cognātus, -a, -um, adj. *kin, related, kindred*
cognōmen, -inis, n. *surname, family name*
colligo, 3, -lēgi, -lēctum, tr. *gather up*
collis, -is, m. *hill, height*
comes, -itis, c. *companion*
comitor, 1, dep. tr. *accompany, go with, attend*
comminus, adv. *hand to hand, at close quarters*
compello, 1, tr. *address, accost*
compōno, 3, -posui, -positum, tr. *place or put together, settle*
cōmptus, -a, -um, p.p.p. as adj. *decked, adorned, ordered*
concēdo, 3, -cessi, -cessum, tr. *grant, hand over*; intr. *withdraw, yield*

VOCABULARY

concipio, 3, -cēpi, -ceptum, tr. *conceive*

concolor, -ōris, adj. *of the same colour*

concurro, 3, -curri, -cursum, intr. *run together, engage in fight*

concutio, 3, -cussi, -cussum, tr. *agitate, shake, rouse*

condēnsus, -a, -um, adj. *crowded, thronged, in thick array*

conditor, -ōris, m. *founder, builder*

condo, 3, -didi, -ditum, tr. *found, hide, shelter*

cōnfectus, see **cōnficio**

cōnficio, 3, -fēci, -fectum, tr. *weaken, wear out, kill*

cōnfugio, 3, -fūgi, intr. *flee for refuge or safety*

congredior, -i, -grēssus, dep. intr. *advance together*

coniungo, 3, -iūnxi, -iūnctum, tr. *join, unite, combine*

coniūnx, -ugis, c. *husband, wife*

conlābor, 3, -lāpsus, dep. intr. *fall down, swoon*

conlāpsus, see **conlābor**

cōnscendo, 3, -ndi, -scēnsum, tr. *climb, mount*

cōnsono, 1, -sonui, intr. *resound, echo*

cōnspicio, 3, -spexi, -spectum, tr. *see, catch sight of*

cōnsūrgo, 3, -surrēxi, -surrēctum, intr. *rise up*

contemno, 3, -tēmpsi, -temptum, tr. *despise, scorn, spurn*

contiguus, -a, -um, adj. *near, within reach of*

contrā, adv. *on the other hand*; prep. with acc. *against*

coorior, 4, -īri, -ortus, dep. intr. *arise, break out*

cor, cordis, n. *heart*

cōram, adv. *face to face*; prep. with abl. *in the presence of, before the eyes of, before*

corniger, -gera, -gerum, adj. *horned*

corōna, -ae, f. *circle, ring, crown*

corpus, -oris, n. *body, strength, person*

corripio, 3, -ripui, -reptum, tr. *snatch, seize, catch*

corruo, 3, -ui, intr. *fall*

costa, -ae, f. *rib*

crāstinus, -a, -um, adj. *of tomorrow, tomorrow's*

crātis, -is, f. *wickerwork, hurdle, framework of a shield*

creātrix, -īcis, f. *mother*

crēdo, 3, -didi, -ditum, tr. with dat. *trust, believe*

Crēsius, -a, -um, adj. *Cretan, of Crete*

crētus, -a, -um, part. as adj. *sprung or descended from*

crīnis, -is, m. *hair, locks*

crūdēlis, -e, adj. *cruel, harsh*

cruentus, -a, -um, adj. *cruel*

cruor, -ōris, m. *blood, gore*

culmen, -inis, n. *roof*

culta, -ōrum, n. pl. *tilled land*

cultus, -ūs, m. *cultivation, civilized way of life*

cum, conj. *when, since*; **cum prīmum**, *as soon as*

cum, prep. with abl. *with*

cumulo, 1, tr. *pile, heap up*

cūnctus, -a, -um, adj., more usually in pl. *all, all in a body*

cupidus, -a, -um, adj. *eager*

VOCABULARY

cupio, 3, -īvi, -ītum, tr. *desire, wish*

cūra, -ae, f. *charge, care, trouble*

cūro, 1, tr. *care for, tend*

currus, -ūs, m. *chariot, car*

cuspis, -idis, f. *spear*

custōs, -ōdis, c. *guard, guardian*

Cyllēnē, -ēs or -ae, f. *Mount Cyllene*

Danaus, -a, -um, adj. *Greek* (gen. pl. **Danaum**)

daps, dapis (no dat. sing. or gen. pl.), f. *feast, banquet*

Dardania, -ae, f. *Troy*

Dardanius, -a, -um, also **Dardanus**, -a, -um, adj. *Trojan*

Dardanus, -i, *son of Jupiter and Electra, ancestor of royal race of Troy*

Daucius, -a, -um, adj. *of or belonging to Daucus*

Daunius, -a, -um, adj. *of or belonging to Daunus, father of Turnus*

Daunus, -i, m. *king of Apulia, father of Turnus*

dē, prep. with abl. *from, down from, of, concerning*

dea, -ae, f. *goddess*

dēbeo, 2, -ui, -itum, tr. *owe, ought*

dēbitus, -a, -um, part. as adj. *due, owed, fitting*

dēcīdo, 3, -cīdi, -cīsum, tr. *cut off, sever*

dēcīsa, see **dēcīdo**

dēcolor, -ōris, adj. *discoloured, tarnished, depraved*

decoro, 1, tr. *honour*

decus, -oris, n. *honour, glory*

dederās, **dedēre**, **dedissem**, see **do**

dēfendo, 3, -di, -sum, tr. *keep, defend*

dēfīgo, 3, -fīxi, -fīxum, tr. *fix*

dēfleo, 2, -flēvi, -flētum, tr. *weep over, lament*

dēfluo, 3, -fluxi, -fluxum, intr. *flow, or float, down*

dehinc, adv. *next, from there*

deinde, adv. *then, next*

dēligo, 3, -lēgi, -lēctum, tr. *choose, pick out*

dēmessum, see **dēmeto**

dēmeto, 3, -messui, -messum, tr. *pluck off, cut off, gather*

dēmitto, 3, -mīsi, -missum, tr. *let down, hang down*

dēmo, 3, dēmpsi, dēmptum, tr. *take, pick, take away*

Dēmodocus, -i, m. *Trojan warrior slain by Halaesus*

dēmoror, 1, dep. tr. *retard, delay, detain*, intr. *linger, tarry*

dēni, distrib. num. *ten each, by tens*

dēnsus, -a, -um, adj. *thick*

dēprecor, 1, dep. tr. *ask mercy*

dērigo, 3, -rēxi, -rēctum, tr. *aim, direct*

dēripio, 3, -ripui, -reptum, tr. *strip off, lop off, tug away*

dēsilio, 4, -silui, -sultum, intr. *leap down, dismount*

dēsisto, 3, -stiti, -stitum, intr. *cease, stand off* (with dat.)

dēspecto, 1, tr. *gaze down on*

dēsum, dēesse, dēfui, intr. *be wanting, fail*

143

VOCABULARY

dēterior, -ius, comp. adj. *worse, degenerate*

deus, -i, m. *god* (gen. pl. **deōrum** or **deum**)

dēvexus, -a, -um, adj. *sloping down, setting, inclining down*

dēvinco, 3, -vīci, -victum, tr. *subdue, conquer;* **dēvicta bella**, *victorious wars*

dexter, -tera (-tra), -terum (-trum), adj. *right*

dextra, -ae, f. *right hand*

dīco, 3, tr. *say, speak, call*

dictum, -i, n. *word, command*

Dīdo, -ūs or -ōnis, f. *Dido, queen of Carthage*

dīdo, 3, -didi, -ditum, tr. *spread abroad*

diēs, diēī, c. *day, dawn, time*

differo, -ferre, distuli, dīlātum, tr. *postpone, put off, delay*

digitus, -i, m. *finger*

dignor, 1, dep. tr. *think worthy*

dignus, -a, -um, adj. *worthy, fit*

dīgressus, -ūs, m. *departure*

dīmitto, 3, -mīsi, -missum, tr. *give up, forgo, send away*

dīruo, 3, -rui, -rutum, tr. *tear apart, wrench away*

dīrus, -a, -um, adj. *dread*

discēdo, 3, -cessi, -cessum, intr. *depart, withdraw, go away*

discerno, 3, -crēvi, -crētum, tr. *pick out, divide, separate*

discrepo, 1, -ui, intr. *differ*

discrīmen, -inis, n. *difference, division, parting, distinction*

discrīmino, 1, tr. *divide, distinguish*

dīsicio, 3, -iēci, -iectum, tr. *scatter, disperse, lay in ruins*

dispergo, 3, -spērsi, -spērsum, tr. *scatter on all sides*

dispērsus, see **dispergo**

diū, adv. *for a long time, long*

divello, 3, -velli, -vulsum, tr. *tear away, separate violently*

dīvus, -i, m. *god, deity* (gen. pl. **dīvum**)

do, dare, dedi, datum, tr. *give, make, cause, grant, utter*

doceo, 2, -ui, doctum, tr. *tell, teach, state, mention*

dolor, -ōris, m. *grief, pain*

domus, -ūs, f. *house, home*

dōnec, conj. *until*

dōnum, -i, n. *gift, present*

dūco, 3, tr. *lead, draw, think, drag on*

ductor, -ōris, m. *chief*

dulcis, -e, adj. *sweet, beloved*

dum, conj. with indic. *while;* with subj. *until, provided that;* **nec dum**, *nor yet*

dūmus, -i, m. *bush, thicket*

duo, -ae, -o, cardinal num. *two*

duplico, 1, tr. *redouble, repeat*

dūro, 1, tr. *harden, undergo,* intr. *endure, suffer*

dūrus, -a, -um, adj. *hard, harsh*

dux, ducis, c. *leader*

ē, ex, prep. with abl. *out of, from*

ecce, interj. *behold, look, see*

ēdo, 3, -didi, -ditum, tr. *give out, put forth, produce, beget*

effātus, see **effor**

effero, -ferre, extuli, ēlātum, irreg. tr. *bring out, bring forth*

effētus, -a, -um, adj. *worn out*

effor, 1, dep. tr. *utter, speak out*

VOCABULARY

effultus, -a, -um, part. as adj. *propped up, supported, couched*

effundo, 3, -fūdi, -fūsum, tr. *pour forth, utter*

egēnus, -a, -um, adj. *poor*

egeo, 2, -ui, intr. *lack, need, be without* (with abl. or, less commonly, gen.)

ego, mei, pron. *I*

ēgredior, 3, -grēssus, dep. intr. *go out or forth, leave*

ēgregius, -a, -um, adj. *distinguished, great, famous*

ēlābor, 3, -lāpsus, dep. intr. *slip away, slip past, escape from*

ēlātam, see **effero**

Ēlectra, -ae, f. *Electra,* daughter of Atlas and mother of Dardanus

ēlīdo, 3, -līsi, -līsum, tr. *strike or force out, crush out*

ēmitto, 3, -mīsi, ēmissum, tr. *send out, throw or hurl out*

emo, 3, ēmi, ēmptum, tr. *buy*

enim, conj. *for*

ēnītor, 3, -nīxus or -nīsus, dep. tr. *bear, give birth to,* intr. *struggle, strive*

ēnīxus, p.p.p. of **ēnītor**

ēnsis, -is, m. *sword, blade*

eo, īre, īvi (ii), **itum,** intr. *go*

epulae, -ārum, f. pl. *feast*

eques, -itis, m. *horseman;* collective sing. *cavalry*

equidem, adv. *certainly, at least, indeed* (usually with first person)

equitātus, -ūs, m. *cavalry*

equus, -i, m. *horse, steed*

ergo, adv. *therefore, for this reason or cause, so then*

erīlis, -e, adj. *of a master*

ēripio, 3, -ripui, -reptum, tr. *snatch away, take away*

error, -ōris, m. *mistake, error*

Erulus, -i, m. *Erulus,* king of Praeneste slain by Evander

et, conj. *and, too, also*

Etrūria, -ae, f. *Etruria*

Etrūscus, -a, -um, adj. *Etruscan*

Euander or **-drus, -dri,** m. Arcadian king, founder of Pallanteum

Eurystheūs, -ei, m. *king of Mycenae in Greece*

Eurytidēs, -ae, m. *son of Eurytus,* Clonus, a goldsmith

ēvincio, 4, -vīnxi, -vīnctum, tr. *bind up or round, wreath*

exanimis (or **exanimus**), **-e,** adj. *lifeless, dead*

exaudio, 4, tr. *hear, give heed to*

excipio, 3, -cēpi, -ceptum, tr. *catch, receive, welcome*

excito, 1, tr. *arouse, awaken, rekindle* (of fire)

exeo, -īre, -ii, -itum, intr. *come out, go out, issue, come forth*

exhortor, 1, dep. tr. *urge on*

exiguus, -a, -um, adj. *small, meagre, weak, thin, slender*

exim, adv. *after that, next*

expedio, 4, tr. *set free, get ready, prepare, put in order, arrange*

exquīro, 3, -sīvi, -sītum, tr. *search out, enquire*

exsors, -sortis, adj. *chosen*

exspecto, 1, tr. *expect, wait for*

exstruo, 3, -strūxi, -strūctum, tr. *pile or heap up, build up*

VOCABULARY

exsul, -ulis, c. *exile, outlaw*
exta, -ōrum, n. pl. *entrails, flesh of sacrifice*
extendo, 3, -tendi, -tēnsum, tr. *stretch out*
externus, -a, -um, adj. *outward, from another country, foreign*
exterreo, 2, tr. *terrify, strike with terror*
extimēsco, 3, -mui, tr. and intr. *fear greatly, be very afraid*
extrēmus, -a, -um, sup. adj. *furthest, last*
extulit, see **effero**
exuo, 3, -ui, -ūtum, tr. *draw off, doff, take off*
exuviae, -ārum, f. pl. *skin, equipment, spoils*

faciēs, -ēi, f. *form, face, appearance*
facilis, -e, adj. *easy, quick*
factum, -i, n. *deed, act*
fāma, -ae, f. *fame, rumour*
famulus, -i, m. *slave, attendant*
fās, indecl. n. *right, proper, lawful*
fastīgium, -i, n. *gable, roof, pinnacle, summit, top*
fātidicus, -a, -um, adj. *prophetic, foretelling fate*
fatīgo, 1, tr. *tire out, torment*
fātum, -i, n. *fate*
Faunus, -i, m. *Faunus, grandson of Saturn; woodland deity, god of the wild country*
faveo, 2, fāvi, fautum, intr. with dat. *favour, support*
fax, facis, f. *torch, flare*
fēlix, -īcis, adj. *fortunate, happy*
ferētrum, -i, n. *litter, bier*
fero, ferre, tuli, lātum, tr. *bear, carry, bring, lead, say, produce*

Ferōnia, -ae, f. *Feronia, old Italian goddess*
ferre, see **fero**
ferrum, -i, n. *iron, iron point, sword*
fervidus, -a, -um, adj. *hot, eager*
fessus, -a, -um, adj. *exhausted*
fētus, -ūs, m. *litter, brood, young*
fidēs, -ei, f. *word of honour, good faith, constancy, pledge, loyalty*
fīdo, 3, fīsus, semi-dep. intr. with dative, *trust, feel confidence*
fīdus, -a, -um, adj. *faithful*
fīgo, 3, fīxi, fīxum, tr. *fix, transfix*
fīlius, -i, m. *son*
fingo, 3, fīnxi, fīctum, tr. *mould, fashion, imagine, suppose, pretend*
fīnis, -is, c. *end, limit, mark*; m. pl. *territories, lands*
fīnitimus, -a, -um, adj. *neighbouring*
firmo, 1, tr. *establish, strengthen*
flamma, -ae, f. *flame, fire*
flecto, 3, flexi, flexum, tr. *bend, turn, persuade*
flexus, -ūs, m. *winding, meander*
flōreo, 2, -ui, intr. *bloom, be bright, flower*
flōs, -ōris, m. *flower, bloom, down on cheek*
flūmen, -inis, n. *river, stream*
fluvius, -i, m. *river, stream*
foedē, adv. *foully, basely*
foedo, 1, tr. *disfigure, mar*
foedus, -eris, n. *covenant, alliance, pact, agreement, treaty*
fōns, fontis, m. *spring*

VOCABULARY

for, 1, dep. tr. *speak, say*
fore, future infin. of **sum**
forēs, -um, f. pl. *door*
fōrma, -ae, f. *beauty, form*
fors, fortis, f. *chance, fate, luck*
forte, adv. *by chance, as it happened*
fortis, -e, adj. *brave, courageous*
Fortūna, -ae, f. *Fortune, Goddess of Fortune*
forum, -i, n. *open space, market-place, market, forum*
fragor, -ōris, m. *crash, thunder*
fremo, 3, -ui, -itum, intr. *murmur, clamour, roar*
frēnum, -i, n. *bridle, rein*
frequēns, -tis, adj. *in crowds, thronging, thick*
frētus, -a, -um, adj. *relying on*
frīgidus, -a, -um, adj. *cold*
frīgus, -oris, n. *cold*
frondōsus, -a, -um, adj. *leafy*
frōns, -ndis, f. *leaf, foliage*
fruor, 3, fructus, dep. intr. with abl. *enjoy, delight in*
frūstrā, adv. *in vain, to no purpose, uselessly, fruitlessly*
fuga, -ae, f. *flight*
fugio, 3, fūgi, fugitum, tr. and intr. *flee, shun, escape*
fulcio, 4, fulsi, fultum, tr. *support, prop up, couch*
fulgeo, 2, fūlsi, intr. *gleam*
fulgor, -ōris, m. *flash of lightning, brightness, gleam*
fultus, p.p.p. of **fulcio**
fulvus, -a, -um, adj. *tawny*
fūmo, 1, intr. *smoke, reek*
fundo, 1, tr. *found, establish*
fundo, 3, fūdi, fūsum, tr. *pour forth, scatter, strew*

fūnereus, -a, -um, adj. *funeral-*
fūnus, -eris, n. *death, funeral*
furiae, -ārum, f. pl. *rage, madness, violent anger* or *passion*
furo, 3, -ui, intr. *rage, rave*
fuscus, -a, -um, adj. *dusky*
futūrus, part. as adj. *future, destined,* see **sum**

galea, -ae, f. *helmet*
gaudeo, 2, gāvīsus, semi-dep. intr. *rejoice, gloat*
gaudium, -i, n. *joy, delight*
gelidus, -a, -um, adj. *icy cold, chilly*
gelus, -ūs, m. *cold, frost, chill*
geminus, -a, -um, adj. *twin-born, paired, two*
gemitus, -ūs, m. *groan, moan*
gemo, 3, -ui, -itum, intr. *groan*
gena, -ae, f. *cheek*
genero, 1, tr. *beget, bring forth*
genitor, -ōris, m. *father, sire, begetter*
gēns, -ntis, f. *nation, race, family*
genus, -eris, n. *descent, race*
glaucus, -a, -um, adj. *grey-green, greyish, bright, gleaming*
globus, -i, m. *band, mass*
glōria, -ae, f. *fame, glory, boasting*
gradior, 3, -i, grēssus, dep. intr. *go, march, proceed*
Grāi, -ōrum, m. pl. *Greeks*
Grāiugena, -ae, m. *Greek by birth, Greek-born* (gen. pl. **Grāiugenum**)
grāmineus, -a, -um, adj. *of grass, grassy, grass-grown*
grandis, -e, adj. *large, great*

VOCABULARY

grātus, -a, -um, adj. *beloved*
gravis, -e, adj. *heavy, hard*
gressus, -ūs, m. *step, course*
grex, gregis, m. *flock, herd, company, crowd*
gutta, -ae, f. *drop, tear*

habeo, 2, tr. *have, hold, keep, consider*
habito, 1, tr. and intr. *live, inhabit, dwell in, haunt*
haereo, 2, haesi, haesum, intr. *cling, stop, halt*
Halaesus, -i, m. Greek settler in Latium, slain by Pallas
harundo, -inis, f. *reed, wreath or crown made of reeds*
haruspex, -icis, m. *augur, soothsayer, prophet*
hasta, -ae, f. *spear*
haud, adv. *not, not at all, by no means*
haurio, 4, hausi, haustum, tr. *draw in, drink up, drain*
hei, interj. *alas!*
Herculeus, -a, -um, adj. *of Hercules*
hērōs, -ōis, m. *hero*
Hēsiona, -ae, f. daughter of Laomedon, king of Troy, mother of Teucer
Hesperia, -ae, f. *land of the west, Italy*
Hesperis, -idis, adj. *of evening, of the west*
hesternus, -a, -um, adj. *of or belonging to yesterday*
hic, haec, hōc, adj. and pron. *this, he, the latter*
hinc, adv. *from here, hence, on this side*

homo, -inis, c. *man, woman*
honor or **honōs**, -ōris, m. *honour, repute, glory*
horreo, 2, intr. *bristle, be tangled or rough, shrink, shudder*
horridus, -a, -um, adj. *rough, shaggy, unkempt, overgrown, dread, dreadful*
hospes, -itis, m. *guest, host, visitor, stranger*
hospitium, -i, n. *bond (between host and guest), friendship*
hostīlis, -e, adj. *of or belonging to an enemy, hostile*
hostis, -is, c. *enemy, foe*
hūc, adv. *hither, to this place*
humilis, -e, adj. *lowly, humble*
humo, 1, tr. *cover with earth*
hyacinthus, -i, m. *iris or lily*
Hȳlaeus, -i, m. a centaur

iaceo, 2, intr. *lie, lie down, lie prostrate*
iacio, 3, iēci, iactum, tr. *throw*
iacto, 1, tr. *throw, hurl, toss*
iam, adv. *now, already, by this time*; **iamdūdum**, *for a long time past, long since*; **iam tum**, *even then*; with negatives, *no longer*
Iāniculum, -i, n. one of the hills of Rome on which Janus was said to have built a town
iānitor, -ōris, m. *doorkeeper*
Iānus, -i, m. *Janus*, old Italian two-headed god
ictus, -ūs, m. *blow*
idem, eadem, idem, pron. *the same*
ignārus, -a, -um, adj. *ignorant*

VOCABULARY

igneus, -a, -um, adj. *fiery, swift as fire*
ignis, -is, m. *fire*
ignōtus, -a, -um, adj. *unknown*
īlex, -icis, f. *holm-oak, ilex*
Īliacus, -a, -um, adj. *Trojan*
Īliades, -um, f. pl. *Trojan women*
ille, illa, illud, dem. pron. *that, he, she, it*
Īlus, -i, m. Latin warrior slain by Pallas
imāgō, -inis, f. *image, shape*
Imāōn, acc. *-ona,* Latin warrior
immānis, -e, adj. *huge, boundless, fierce*
immātūrus, -a, -um, adj. *too soon, too early, unripe, untimely*
immolo, 1, tr. *slay*
immūgio, 4, -īvi (-ii), intr. *resound, re-echo, roar*
impār, -aris, adj. *unequal*
impello, 3, -puli, -pulsum, tr. *drive or push on, urge*
imperditus, -a, -um, adj. *undestroyed, not slain*
imperium, -ii, n. *command, sway, realm, dominion*
impero, 1, tr. and intr. with dat. *command, order*
impōno, 3, -posui, -positum, tr. *lay on, place on*
impressum, see **imprimo**
imprimo, 3, -pressi, -pressum, tr. *press or stamp on*
imprōvīsō, adv. *unexpectedly*
īmus, -a, -um, adj. *very deep, lowest, deepest*
in, prep. with acc. *into, to, against;* with abl. *in, on*

inānis, -e, adj. *vain, empty, useless*
incautus, -a, -um, adj. *heedless*
incendium, -i, n. *fire*
incendo, 3, -di, -sum, tr. *set fire to, kindle, burn*
inceptum, -i, n. *attempt, undertaking*
incertus, -a, -um, adj. *uncertain, doubtful, unsure*
incesto, 1, tr. *dishonour*
incido, 3, -cidi, -cāsum, intr. *fall on, come up, meet*
incipio, 3, -cēpi, -ceptum, tr. *begin, start*
inclūdo, 3, -ūsi, -ūsum, tr. *shut in, close in*
incolo, 3, -ui, -cultum, tr. *inhabit, live in, dwell in*
incolumis, -e, adj. *safe*
incommodus, -a, -um, adj. *unsuitable, troublesome, disagreeable;* **incommodum, -i,** n. *trouble, misfortune*
increpo, 1, -ui, -itum, intr. *resound, re-echo*
incumbo, 3, -cubui, -cubitum, intr. *lean forward, lean on*
inde, adv. *then, from there*
indignor, 1, dep. tr. and intr. *be indignant, be displeased, consider unworthy*
indiscrētus, -a -um, adj. *indistinguishable*
indocilis, -e, adj. *untaught, difficult to teach*
indulgeo, 2, -dūlsi, -dultum, tr. and intr. *be kind, tender towards* (with dat.)
induo, 3, -ui, -ūtum, tr. *clothe*
inēluctābilis, -e, adj. *inevitable, overpowering, inescapable*

VOCABULARY

inermus (also **inermis**, -e), -a, -um, adj. *unarmed*

inexplētus, -a, -um, adj. *unsatisfied, insatiable, not filled*

infandus, -a, -um, adj. *unspeakable, unutterable, horrible*

infēlix, -īcis, adj. *unhappy*

inferiae, -ārum, f. pl. *offerings or sacrifices to the dead*

infero, -ferre, -tuli, illātum, tr. *bring upon, bear against*; **sē inferre**, *enter, bear down upon*

infrā, adv. and prep. with acc. *below*

ingēns, -ntis, adj. *great, vast*

ingredior, 3, -i, -grēssus, dep. tr. *attempt, enter*

ingruo, 3, -ui, intr. *rush or break into, fall violently upon*

inhaereo, 2, -haesi, -haesum, intr. *cling or cleave to, stick to*

inicio, 3, -iēci, -iectum, tr. *throw upon, cast upon, inspire*

inimīcus, -a, -um, adj. *of an enemy, hostile, unfriendly*

iniquus, -a, -um, adj. *uneven, sloping, unfair, unjust*

inlacrimo, 1, tr. *weep over*

inmitto, 3, -mīsi, -missum, tr. *throw, hurl upon, loosen*

inno, 1, intr. *float on, swim on*

inopīnus, -a, -um, adj. *unexpected, unforeseen*

inops, -opis, adj. *without resources, scanty, weak, poor*

inquam, inquis, inquit, defect. *say*

inreparābilis, -e, adj. *that cannot be renewed or recovered*

insequor, 3, -i, -secūtus, dep. tr. *follow after, pursue*

insideo, 2, -sēdi, -sessum, tr. and intr. *settle, occupy*

insignia, -ium, n. pl. *badges or marks of honour*

insignis, -e, adj. *distinguished*

instauro, 1, tr. *repeat, renew (of ritual)*

insto, 1, -stiti, —, intr. *press on, come near, threaten, menace*

instruo, -ere, -xi, -ctum, tr. *draw up in order, equip, make ready*

insuētus, -a, -um, adj. *unaccustomed, strange, unwonted*

insulto, 1, intr. *mock, exult*

intāctus, -a, -um, adj. *untouched, unhurt*

inter, prep. with acc. *among, between, during, amid*

intercipio, 3, -cēpi, -ceptum, tr. *catch between, intercept*

intereā, adv. *meanwhile, in the meantime*

interimo, 3, -ēmi, -ēmptum, tr. *slay*

intersum, -esse, -fui, intr. *take part in, be present at* (with dat.)

intertexo, 3, -texui, -textum, tr. *interweave*

intorqueo, 2, -torsi, -tortum, tr. *hurl against*

inumbro, 1, tr. *cover, shadow*

invenio, 4, -vēni, -ventum, tr. *come upon, find, find out*

invictus, -a, -um, adj. *unconquered, unconquerable*

invideo, 2, -vīdi, -vīsum, intr. with dat. *envy, grudge*

inviso, 3, -visi, visum, tr. *visit, see*

invīto, 1, tr. *invite*

VOCABULARY

ipse, ipsa, ipsum, pron. and adj. *-self, the very*

ira, -ae, f. *anger, wrath, rage*

is, ea, id, adj. and pron. *that; he*

iste, ista, istud, dem. pron. *that of yours, that*

Italus, -a, -um, adj. *Italian*

iter, itineris, n. *road, way, course, journey*

iterum, adv. *a second time, again*

iubeo, 2, iūssi, iūssum [old fut. iūsso], tr. *bid, order*

iugālis, -e, adj. *nuptial, wedding-*

iugulum, -i, n. *throat, gullet*

iugum, -i, n. *yoke, height, ridge*

Iūlus, -i, m. *son of Aeneas*

iungo, 3, iūnxi, iūnctum, tr. *join, attach, unite*

Iūno, -ōnis, f. *wife of Jupiter*

Iuppiter, Iovis, m. *Jupiter, father of the gods*

iūssum, -i, n. *order, command*

iūstus, -a, -um, adj. *just, fair*

iuvenis, -is, adj. *young*; noun, c. *young man, young girl or woman, youth, warrior*

iuventās, -ātis, f. *youth*

iuventūs, -ūtis, f. *youth*

iuvo, 1, iūvi, iūtum, tr. *help*; [often impersonal] *it delights, it comforts, it pleases*

iūxtā, adv. *hard by, near by, nigh*

labor, -ōris, m. *toil, trouble*

lābor, 3, -i, lāpsus, dep. intr. *glide, glide away, slip, droop, fall*

labōro, 1, tr. and intr. *work, be in difficulties, be afflicted*

lacrima, -ae, f. *tear*

lacrimo, 1, intr. *weep*

lacus, -ūs, m. *hollow, pool, lake*

Lādōn, -ōnis, m. Trojan warrior slain by Halaesus

laetus, -a,-um, adj. *happy*

laevus, -a, -um, adj. *left*

langueo, 2, -ui, intr. *droop*

lānx, lancis, f. *dish, platter*

Lāomedontiadēs, -ae, m. *descendant of Laomedon,* Priam, king of Troy

lār, laris, m. *hearth, dwelling-place*

largior, 4, -īri, -ītus, dep. tr. *give, grant*

Lārīdēs, -ae, m. Latin warrior, twin-brother of Thymber

lātē, adv. *far and wide*

lateo, 2, -ui, intr. *lie hid, lie in hiding, lurk*

Latīnus, -i, m. king of Latium

Latīnus, -a, -um, adj. *of or belonging to Latium, Latin*

Latium, -i, n. district round and south of Rome

lātūram, see **fero**

latus, -eris, n. *side, flank*

lātus, -a, -um, adj. *wide*

laudo, 1, tr. *praise*

Laurēns, -ntis, adj. *Laurentian, of or belonging to Laurentum*

laus, laudis, f. *praise, honour*

lautus, -a, -um, part. as adj. *splendid, sumptuous, neat*

laxo, 1, tr. *open, unloose, slacken*

lēgātus, -i, m. *ambassador, envoy, representative, messenger*

legio, -ōnis, f. *legion*

lego, 3, lēgi, lēctum, tr. *read, choose, pick up, trace, collect*

Lēmnius, -a, -um, adj. *of Lemnos*

VOCABULARY

lēnio, 4, tr. *calm, soothe*
leo, -ōnis, m. *lion*
Lernaeus, -a, -um, adj, *of Lerna, a forest and marsh near Argos*
lētum, -i, n. *death*
lēvis, -e, adj. *smooth*
levo, 1, tr. *lighten, relieve, cheer*
libēns, -entis, adj. *willing, glad*
libro, 1, tr. *balance, poise, aim*
Libystis, -idis, f. *Libyan*
licet, 2, licuit, impersonal, *it is allowed*
licitus, -a, -um, part. as adj. *permitted, allowable, lawful*
līmen, -inis, n. *threshold, door*
litoreus, -a, -um, adj. *of or belonging to the sea-shore, shore-, beach-*
lītus, -oris, n. *shore, beach, bank*
loco, 1, tr. *place, set*
locus, -i, m. (pl. loci, or loca, n.), *place*
longaevus, -a, -um, adj. *aged*
longē, adv. *far, from afar, at length*
longus, -a, -um, adj. *long, deep*
loquor, 3, -i, locūtus, dep. tr. and intr. *speak, say*
lōrīca, -ae, f. *corselet*
lūceo, 2, -xi, intr. *shine, gleam*
Lūcifer, -eri, m. *morning star (light-bringing)*
luctāmen, -inis, n. *struggle*
lūctus, -ūs, m. *grief, sorrow*
lūcus, -i, m. *grove*
lūmen, -inis, n. *light, eye, glory;* lūmina ducum, *glorious leaders*
Lupercal, -ālis, n. *grotto on the Palatine Hill*

lūstrālis, -e, adj. *belonging to a sacrifice of purification, sacrificial*
lūx, lūcis, f. *light, daylight*
Lycaeus, -a, -um, adj. *of Lycaeus (a mountain in Arcadia)*
Lycius, -a, -um, adj. *Lycian*
Lȳdius, -a, -um, adj. *of or belonging to Lydia (a district in Asia Minor)*

macto, 1, tr. *sacrifice*
Maeonia, -ae, f. *Maeonia, Lydia*
Maeonius, -a, -um. adj. *of or belonging to Maeonia*
maestus, -a, -um, adj. *sad*
mage, adv. *more*
magister, -tri, m. *master*
magnus, -a, -um, adj. *great*
Māia, -ae, f. *Maia, daughter of Atlas, mother of Mercury*
mālo, mālle, mālui, tr. *prefer*
mandātum, -i, n. *command*
mando, 1, tr. *commit to someone's charge, consign, commend*
maneo, 2, mānsi, mānsum, tr. or intr. *wait, abide, await, remain*
mānēs, -ium, m. pl. *shades, spirits of the dead*
manus, -ūs, f. *hand, might, labour, band, throng*
mare, -is, n. *sea*
Mārs, Mārtis, m. *Mars, god of war, war, warfare*
māter, -tris, f. *mother*
mātūtīnus, -a, -um, adj. *of morning, morning*
maximus, -a, -um, sup. adj. *greatest, very great*

152

VOCABULARY

meditor, 1, dep. tr. *meditate, plan, intend*; intr. *brood*

medius, -a, -um, adj. *middle of, midst of*; **in mediōs**, *into the midst*

membrum, -i, n. *limb*

memini, -isse, defect. *remember*

memor, -oris, adj. *mindful*

memoro, 1, tr. *speak of, relate, remind, say*

mēns, -tis, f. *mind, will, intent*

mēnsa, -ae, f. *table*

Mercurius, -i, m. *Mercury, son of Jupiter and Maia, messenger of the gods*

mereo, 2, -ui, -itum, tr. *deserve, win, earn*

meritum, -i, n. *merit, desert*

meritus, -a, -um, adj. *deserved*

mēta, -ae, f. *goal*

metuo, 3, -ui, -ūtum, tr. *fear*

metus, -ūs, m. *fear, panic*

meus, -a, -um, poss. adj. *my, mine*

Mezentius, -i, m. *an Etruscan leader*

mīlitia, -ae, f. *war, warfare*

mīlle, num., pl. **mīlia**, *thousand* (poetic, *many*)

minae, -ārum, f. pl. *threats*

ministro, 1, tr. *serve, supply*

minor, 1, dep. tr. *threaten*

minus, comp. adv. *less*

mīrābilis, -e, adj. *wonderful*

mīror, 1, dep. tr. and intr. *wonder at, admire, be astonished*

misceo, 2, mīscui, mīxtum, tr. *mix, mingle*

miser, -era, -erum, adj. *wretched, unhappy, luckless, poor*

miserābilis, -e, adj. *pitiable*

miserandus, -a, -um, gerundive as adj. *pitiable, piteous*

miserēsco, 3, intr. *feel pity* (with gen.)

miseror, 1, -ātus, dep. tr. *pity*

missilis, -e, adj. *for hurling or throwing*; **-is**, n. *missile, a weapon which is thrown*

mītis, -e, adj. *kind, gentle, smooth*

mitto, 3, mīsi, missum, tr. *send, pass over, omit*

mŏdŏ, adv. *lately*

modus, -i, m. *fashion, way, due measure, manner*

moenia, -ium, n. pl. *walls, ramparts, walled town*

mōlior, 4, -ītus, dep. tr. and intr. *work, toil, exert oneself*

mollis, -e, adj. *soft, tender*

moneo, 2, -ui, -itum, tr. *warn, advise, bid*

monimentum, -i, n. *monument, memorial, record*

monitum, -i, n. *advice, warning*

monitus, -ūs, m. *bidding*

mōns, -ntis, m. *mountain, height*

mōnstro, 1, tr. *show, point out*

mōnstrum, -i, n. *monster, portent, strange appearance*

mora, -ae, f. *delay, lingering*

morior, 3, mori, mortuus, dep intr. *die*

moror, 1, dep. tr. *prolong, impede, delay*, intr. *linger*

mors, -tis, f. *death*

mortālis, -e, adj. *mortal*

mōs, mōris, m. *custom*; pl. *character*; **in mōrem**, *like*

moveo, 2, mōvi, mōtum, tr. *move, trouble, wield, disturb*

VOCABULARY

mox, adv. *soon*
mūgio, 4, -īvi (-ii), -ītum, intr. *low, bellow*
multus, -a, -um, adj. *much, many* [comp. **plūs**; sup. **plūrimus**]
mūnus, -eris, n. *gift, boon, favour, duty, prize*
mūrus, -i, m. *wall*

nam, namque, conj. *for*
nāscor, 3, -i, nātus, dep. intr. *be born, be sprung from*
nātūra, -ae, f. *nature, character*
nātus, -i, m. *son*
nāvis, -is, f. *ship, boat*
-ne, enclitic interrog. particle
nē, conj. *not*; with subjunctive, *do not, that...not, lest*; with imperative, *not* (poetic usage)
nec, neque, conj. *nor, and not*; **nec...nec**, *neither...nor*; **nec nōn**, *and*
nefās, indecl. n. *disgrace, wrong, sin, crime*
nego, 1, tr. *deny, refuse*
Nemea, -ae, f. *Nemea*, city in Argolis in Greece
Nemeus, -a, -um, adj. *of Nemea*
nemus, -oris, n. *grove, wood*
neque, see **nec**
nescius, -a, -um, adj. *ignorant*
neu, nēve, conj. *and...not* (= et nē)
nī, conj. *unless, if...not*
niger, -gra, -grum, adj. *black*
nigrāns, -antis, part. as adj. *dusky, dark*
nihil, nīl, n. indecl. *nothing*; as adv. *not at all*

nimbus, -i, m. *rain-cloud, thunder-cloud*
niveus, -a, -um, adj. *snow-white, pale*
nōbilis, -e, adj. *noble, famous*
nōbiscum, see **nōs**
nōdus, -i, m. *knot*
nōmen, -inis, n. *name, reputation*
nōn, adv. *not*; **nec nōn**, *and*
nōs, nostri or nostrum, pl. pron. *we, I*
nōsco, 3, nōvi, nōtum, tr. *get to know, learn*; p.p.p. **nōtus**, -a, -um, *known*
noverca, -ae, f. *stepmother*
novus, -a, -um, adj. *new*
nox, noctis, f. *night, darkness*
nūbēs, -is, f. *cloud*
nūbigena, -ae, c. *cloud-born*
nūllus, -a, -um, adj. *no one, no*
num, interrog. particle, *whether, surely...not*
nūmen, -inis, n. *will, divine will, divine power*
nunc, adv. *now*; **nunc... nunc**, *now...then, at one time...at another*
nunquam, adv. *never*
nūntia, -ae, f. *messenger*
nūntius, -i, m. *message, news, messenger*
Nympha, -ae, f. *nymph*, spirit of waters, woods, mountains, goddess of the countryside

ō, interj. *O, oh!*
obeo, -īre, -īvi (-ii), -itum, tr. *go against, overspread, cover*
ōbex, -icis, c. *bolt, barrier*
ōbicio, 3, -iēci, -iectum, tr. *throw before, throw in way of*

154

VOCABULARY

obnūbo, 3, nūpsi, nūptum, tr. *veil, shroud*

oborior, 4, -īri, -ortus, dep. intr. *arise*

obruo, 3, -rui, -rutum, tr. *overwhelm*

obsitus, -a, -um, part. as adj. *covered over, weighed down*

obstipēsco, 3, -stipui, —, intr. *be amazed*

obtentus, -ūs, m. *covering, spread*

obtrunco, 1, tr. *cut down*

obvius, -a, -um, adj. *coming to meet, meeting*

occido, 3, -cidi, -cāsum, intr. *fall, perish*

occīdo, 3, -cīdi, -cīsum, tr. *kill*

Ōceanus, -i, m. *Ocean*

ōcior, -us, comp. adj. *swifter*

ōcius, comp. adv. *more swiftly*

oculus, -i, m. *eye*

ōdi, ōdisse, defect. tr. *hate*

odium, -i, n. *hate, hatred, spite*

Oechalia, -ae, f. *city in Euboea in Greece*

ōlim, adv. *formerly, once*

oliva, -ae, f. *olive, olive tree*

olli, old form of *illi*

Olympus, -i, m. *a mountain, home of gods, heaven*

omnipotēns, -tis, adj. *all-powerful, almighty*

omnis, -e, adj. *all, the whole, every*

onero, 1, tr. *load, heap up*

opācus, -a, -um, adj. *thick*

opem (no nom.), -is, f. *help, aid, power*; pl. *wealth, resources*

opīmus, -a, -um, adj. *rich*

oppidum, -i, n. *town*

oppōno, 3, -posui, -positum, tr. *expose, offer, set against*

oppositus, see **oppōno**

optimus, -a, -um, sup. adj. *best, excellent, very great*

opto, 1, tr. *wish, long for, desire*

opulentus, -a, -um, adj. *wealthy, rich, powerful*

opus, -eris, n. *work, toil, labour*

ōra, -ae, f. *shore, edge, rim*

ōrāculum, -i, n. *oracle, prophecy*

ōrātor, -ōris, m. *speaker*

orbis, -is, m. *circle, world*

Orcus, -i, m. *the lower world, the abode of the dead, Orcus*

ōrdo, -inis, m. *order, line, rank*

orior, 4, orīri, ortus, dep. intr. *arise, rise*

ōro, 1, tr. and intr. *pray*

ōs, ōris, n. *mouth, tongue, voice*; pl. *face*

ŏs, ossis, n. *bone*

ostento, 1, tr. *show, point out*

ostrum, -i, n. *purple, crimson*

ovāns, -tis, part. as adj. *triumphant, exulting, joyful*

pācifer, -era, -erum, adj. *peace-bringing, peaceful*

Pallantēum, -i, n. *city founded by Evander*

Pallās, -ntis, m. *son of Evander*

palma, -ae, f. *hand, palm of the hand*

palūs, -ūdis, f. *marsh, pool*

Pān, Pānos, m. *god of woods and shepherds*

pango, 3, pepigi, pāctum, tr. *fix, covenant*

panthēra, -ae, f. *panther*

pār, paris, adj. *equal, like*

VOCABULARY

Parcae, -ārum, f. pl. *the Fates*
parco, 3, peperci, parsum, intr. with dat. *spare*
parēns, -ntis, c. *parent*
pario, 3, peperi, partum, tr. *get, produce, beget*
pariter, adv. *likewise, in the same manner, equally*
paro, 1, tr. *prepare, make ready*
Parrhasius, -a, -um, adj. *Arcadian*
pars, -tis, f. *part*
partus, -a, -um, part. as adj. *gained, acquired*
parvus, -a, -um, adj. *small, little*
pāssim, adv. *far and wide*
pateo, 2, -ui, intr. *lie open, be exposed*
pater, -tris, m. *father*
patior, 3, passus, dep. tr. *suffer, endure, undergo*
patria, -ae, f. *country, native land*
patrius, -a, -um, adj. *of one's country, native, ancestral*
pauci, -ae, -a, pl. adj. *few*
paulātim, adv. *little by little*
pauper, -eris, adj. *poor*
pavidus, -a, -um, adj. *terrified*
pāx, pācis, f. *peace*
pectus, -oris, n., also pl., *breast, chest, heart*
pecus, -oris, n. *cattle, flock, herd*
pecus, -udis, f. *beast, single head of cattle*; pl. *cattle*
pedes, -itis, m. *foot-soldier*; adj. *on foot*
pedester, -tris, -tre, adj. *on foot, foot-*
pelagus, -i, n. *sea, ocean*
Pelasgi, -ōrum, m. pl. *Pelasgians, oldest inhabitants of Greece*
pellis, -is, f. *skin, hide, pelt*
pello, 3, pepuli, pulsum, tr. *drive, drive out, rout, vanquish*
penātēs, -ium, m. pl. *household gods, home, house*
penetrābilis, -e, adj. *piercing*
penitus, adv. *deeply, utterly*
pepigi, see **pango**
per, prep. with acc. *through, throughout, over, amongst, on*
percutio, 3, -cussi, -cussum, tr. *strike through, astound*
perdo, 3, -didi, -ditum, tr. *lose, waste*
perfero, -ferre, -tuli, -lātum, tr. *bring, carry, bear through*
perficio, 3, -fēci, -fectum, tr. *finish, perfect*
perfundo, 3, -fūdi, -fūsum, tr. *bedew, bathe, dip in*
perfūsōs, see **perfundo**
Pergama, -ōrum, n. pl. *the citadel of Troy, Troy*
perhibeo, 2, -ui, -itum, tr. *hold out, extend, attribute, ascribe, say, assert, name*
perīculum (also **perīclum**), -i, n. *danger, peril, hazard, risk*
permīsceo, 2, -mīscui, -mīxtum, tr. *mix, mingle*
perpetuus, -a, -um, adj. *unbroken, continuous*
persolvo, 3, -vi, -solūtum, tr. *offer in payment, offer, pay*
pertulerit, see **perfero**
pēs, pedis, m. *foot*
peto, 3, -īvi (-ii), -ītum, tr. *ask, ask for, make for, aim at, seek, look for, attack*

VOCABULARY

phalanx, -ngis, f. *band, troop*
pharetra, -ae, f. *quiver*
Pheneus, -i, f. *a town in Arcadia*
Pherēs, -ētis, m. *Trojan warrior slain by Halaesus*
Pholus, -i, m. *a centaur*
Phryges, -um, m. pl. *Phrygians*
Phrygius, -a, -um, adj. *Phrygian*
pīctus, -a, -um, p.p.p. as adj. *coloured, painted, decorated*
pingo, 3, pīnxi, pīctum, tr. *paint, embroider*
pinguis, -e, adj. *fat, rich*
pius, -a, -um, adj. *dutiful*
placidus, -a, -um, adj. *quiet*
plango, 3, plānxi, plānctum, tr. *beat the breast*, intr. *wail*
planta, -ae, f. *sole, foot*
plēnus, -a, -um, adj. *full*
plūs, plūris, comp. adj. *more*
pōculum, -i, n. *cup, goblet*
poena, -ae, f. *penalty, punishment*
pollex, -icis, m. *thumb*
pompa, -ae, f. *solemn procession*
pondus, -eris, n. *weight*
pōno, 3, posui, positum, tr. *place, put, set, offer, found, pitch* (camp)
pontus, -i, m. *sea*
pōpuleus, -a, -um, adj. *of or belonging to poplars, poplar-*
pŏpulus, -i, m. *people, nation*
porta, -ae, f. *gate, gateway*
portentum, -i, n. *portent, sign*
pōsco, 3, popōsci, tr. *ask*
post, adv. *behind, afterwards*; prep. with acc. *behind, after*
postquam, conj. *after*
postrēmus, -a, -um, sup. adj. *last*

potēns, -ntis, adj. *powerful*
potior, 4, -īri, -ītus, dep. intr. *gain, become master of* (with abl.)
potis, -e, adj. *able, possible*
Potītius, -ii, m. *name of a Roman priestly family*
potītus, see potior
praecēdo, 3, -cessi, -cessum, intr. *go before, precede*
praecipuus, -a, -um, adj. *distinguished, conspicuous, chief*
praeclārus, -a, -um, adj. *famous, renowned, distinguished*
praecordia, -ōrum, n. pl. *heart, breast*
praeda, -ae, f. *spoil, booty*
praedulcis, -e, adj. *very sweet*
praefīgo, 3, -fīxi, -fīxum, tr. *fix in front, tip*
praefulgeo, 2, -fūlsi, intr. *shine forth, glitter before* or *in front of*
praemium, -i, n. *reward, booty*
praenūntius, -a, -um, adj. *foretelling, heralding*
praesēns, -entis, adj. *present, immediate, instant*
praesidium, -i, n. *protection*
praestāns, -ntis, part. as adj. *outstanding, surpassing*
praetendo, 3, -di, -tum, tr. *hold before, stretch forth*
praeter, adv. *past, beyond*
praeteritus, -a, -um, part. as adj. *gone by, bygone, departed*
precor, 1, dep. tr. *pray, pray to, entreat, beseech*
premo, 3, pressi, pressum, tr. *press, keep down, draw tight*
(prex), precis, f. *prayer, petition*; usually pl. **preces, precum**

157

VOCABULARY

Priamus, -i, m. *Priam, king of Troy*
primitiae, -ārum, f. pl. *first-fruits*
primus, -a, -um, adj. *first, earliest, foremost, chief*
prior, prius, comp. adj. *earlier, former, first, quicker, better*
priscus, -a, -um, adj. *ancient*
proavus, -i, m. *great-grandfather*
prōcēdo, 3, -cessi, -cessum, intr. *go forward*
procerēs, -um, m. pl. *chiefs*
procul, adv. *far, afar*
prōcumbo, 3, -cubui, -cubitum, intr. *fall down, lie down, lean forward*
prōdigium, -i, n. *portent*
proelium, -i, n. *battle, combat*
profectō, adv. *indeed*
profectus, p.p. of **proficiscor**
proficīscor, -i, -fectus, dep. intr. *set out*
profugus, -a, -um, adj. *fugitive, exiled, outlaw*
prōgeniēs, -ēi, f. *offspring*
prōgredior, 3, -grēssus, dep. intr. *advance, go forward*
prōicio, 3, -iēci, -iectum, tr. *cast down, cast forth*
prōiectus, see **prōicio**
prōlēs, -is, f. *offspring, child*
prōmissum, -i, n. *promise*
prōmitto, 3, -mīsi, -missum, tr. *promise*
prōnus, -a, -um, adj. *hanging forward, bending down*
prope, prep. with acc. *near, nigh*; adv. *nearby*; comp. adj.
propior, -ius, *nearer*
propero, 1, intr. *hasten, hurry*

propinquo, 1, intr. *draw near*
propinquus, -a, -um, adj. *near, neighbouring, near at hand*
prōra, -ae, f. *prow*
prōrumpo, 3, -rūpi, -ruptum, intr. *break out, rush out or forward, charge*, tr. *fling forth*
prōtendo, 3, -di, -sum or -tum, tr. *stretch forward*
prōtinus, adv. *straightway, at once, immediately*
pūbēs, -is, f. *youth*; collective, *young men*
pudendus, -a, -um, gerundive as adj. *shameful, disgraceful*
pudor, -ōris, m. *shame, modesty, good manners, sense of duty*
pugna, -ae, f. *battle, fight*
pugnus, -i, m. *fist*
pulcher, -chra, -chrum, adj. *beautiful* (superl. **pulcherrimus**)
pulmo, -ōnis, m. *lung*
pulvereus, -a, -um, adj. *dusty*
puppis, -is, f. *stern, ship*
puto, 1, tr. *think, ponder*
putris, -tris, -tre, adj. *rotten, crumbling, dusty*

quā, adv. *where*
quadripedāns, -ntis, adj. *four-footed, galloping*
quaero, 3, -sīvi, -sītum, tr. *seek, ask*
quaeso, old form of **quaero**
quam, adv. *than, as, how*
quandō, conj. *since, because*; adv. *ever*
quantus, -a, -um, adj. *how great, as great as*

VOCABULARY

quatio, 3, —, quassum, tr. *shake, harry*

-que, enclitic particle, *and*; **-que...-que**, *both...and*

quercus, -ūs, f. *oak, oak-tree*

quernus, -a, -um, adj. *of oak*

quī, quae, quod, rel. pron. *who, which*

quīcumque, quae-, quod-, indef. pron. *whoever, whosoever, whatsoever*

quidem, adv. *indeed*

quiēs, -ētis, f. *rest, repose, quiet*

quīn, adv. *why even*; conj. *but that, that...not*

quisquam, quae-, quid-(quic-), indef. pron. *any (one or thing)*

quisque, quae-, quic-, pron. *each*

quisquis, quicquid, pron. *whoever*

quīvīs, quae-, quod-, indef. pron. *any you please, whatever*

quō, adv. *whither? to what end?*

quod, conj. (properly n. sing. of quī), *that, because*; **quod sī**, *but if*

quondam, adv. *once, formerly*

quoniam, conj. *since, because*

quoque, conj. *also, too*

rabiēs, -em, -e, f. *madness, rage*

rāmus, -i, m. *branch, bough*

rapio, 3, -ui, raptum, tr. *snatch up, seize and carry off*

rārus, -a, -um, adj. *thin, scattered, here and there*

ratio, -ōnis, f. *thought, plan*

ratis, -is, f. *raft, ship*

recēdo, 3, -cessi, -cessum, intr. *go back, depart*

recepto, 1, tr. *recover, get back*

recipio, 3, -cēpi, -ceptum, tr. *retrace, receive, take back*

recondo, 3, -didi, -ditum, tr. *hide, bury*

recordor, 1, dep. tr. *remember*

rector, -ōris, m. *ruler*

rēctus, -a, -um, adj. *straight, right*

recubo, 1, intr. *lie back*

reddo, 3, -didi, -ditum, tr. *give back, make, render, bring back*

redeo, -īre, -ii, -itum, intr. *return, come back*

reditus, -ūs, m. *return*

refero, -ferre, rettuli, -lātum, tr. *bring back, give back, relate, say, withdraw, restore*; **sē referre**, *return*

regio, -ōnis, f. *district, line, direction, quarter*

rēgnum, -i, n. *kingdom*

rego, 3, rēxi, rēctum, tr. *rule, guide*

rēicio, 3, -iēci, -iectum, tr. *throw back, beat back*

religio, -ōnis, f. *religious awe, sanctity, scruple, rite, worship*

relinquo, 3, -līqui, -lictum, tr. *leave, forsake, abandon, desert*

rěliquiae, -ārum, f. pl. *relics*

rēmigium, -ii, n. *rowing, oars*

remitto, 3, -mīsi, -missum, tr. *send back*

rēmus, -i, m. *oar*

repente, adv. *suddenly*

repleo, 2, -plēvi, -plētum, tr. *fill*

repōno, 3, -posui, -positum, tr. *set down, restore, put back*

VOCABULARY

repōsco, 3, tr. *demand, demand or ask back*
reprimo, 3, -pressi, -pressum, tr. *check, repress, hold back*
requiēs, -ētis, f. *rest, respite*
rēs, rĕi, f. *thing, matter, task*; pl. *fortunes*
reservo, 1, tr. *keep back, preserve*
resīdo, 3, -sēdi, intr. *sit down*
resolvo, 3, -solvi, -solūtum, tr. *loosen, disperse*
resto, 1, -stiti, intr. *remain*
resulto, 1, intr. *rebound, resound*
retineo, 2, -tinui, -tentum, tr. *hold back, restrain, detain*
retorqueo, 2, -torsi, -tortum, tr. *twist or fling back*
retracto, 1, tr. *draw back*
reveho, 3, -vĕxi, -vectum, tr. *bring back, carry back*
revīso, 3, tr. and intr. *visit again, come to see again*
rēx, rēgis, m. *king*
Rhoetus, -i, m. king of the Marruvii, a race who lived in the Apennines
rigeo, 2, intr. *be stiff, stiffen*
rīpa, -ae, f. *bank*
rīte, adv. *fitly, duly, rightly*
rōbur, -oris, n. *oak, oak-wood, vigour, strength, shaft of oak-wood*
rogo, 1, tr. *ask, request*
Rōmānus, -a, -um, adj. *Roman*
Rōmulus, -i, m. *Romulus*, founder and first king of Rome
roto, 1, tr. *wheel, turn, revolve*
rudīmentum, -i, n. *beginning*
rūmor, -ōris, m. *noise, rumour*

rumpo, 3, rūpi, ruptum, tr. *break, burst open, interrupt*; **sē rumpere**, *burst forth*
ruo, 3, rui, rutum, intr. *rush, go to ruin, fall, rush down*
rūpēs, -is, f. *rock, cliff, crag*
rutilo, 1, intr. *redden, have a reddish glow*
Rutulus, -i, m. *Rutulian*, man of Ardea

Sabellus, -a, -um, adj. *Sabellian, Samnite*
sacer, sacra, sacrum, adj. *sacred, holy*; n. pl. **sacra**, -ōrum, *sacred rites, sacred objects, a sacrifice*
sacerdōs, -ōtis, c. *priest, priestess*
saeculum, -i, n. *generation, age*
saepe, adv. *often*; comp. **saepius**
saevus, -a, -um, adj. *cruel*
sagitta, -ae, f. *arrow*
Salamīs, -mīnis, f. island of Salamis
Salii, -ōrum, m. pl. a college of dancing priests sacred to Mars
salūs, -ūtis, f. *safety, health*
salvē, see **salveo**
salveo, 2, intr. *be well*; imperative, **salvē**, *hail! greeting!*
sānctus, -a, -um, adj. *sacred*
sānguis, -inis, m. *blood*
saniēs, -em, -ē, f. *venom*
Sarpēdōn, -onis, m. *Sarpedon*, a Trojan warrior
Sāturnius, -a, -um, adj. *of or belonging to Saturn*
Sāturnus, -i, m. *Saturn*, ancient

VOCABULARY

king of Latium, god of agriculture
satus, -a, -um, p.p.p. of **sero**, 3, sēvi, satum, *sow*; with abl. *son of, sprung from*
saxum, -i, n. *rock, stone*
scelero, 1, tr. *defile, pollute*; p.p.p. **scelerātus**, *wicked*
scēptrum, -i, n. *sceptre*
scindo, 3, scidi, scīssum, tr. *split, cleave, branch off*
scūtum, -i, n. *shield*
sē, or **sēsē**, sui, reflexive pron. *him-, her-, itself*; pl. *themselves*
seco, 1, -ui, sectum, tr. *cut*
sēcrētum, -i, n. *solitary place*
secundus, -a, -um, adj. *second, favourable, following*
sed, conj. *but*
sedeo, 2, sēdi, sēssum, intr. *sit, sit still, be fixed*
sēdēs, -is, f. *home, abode, dwelling*
sedīle, -is, n. *seat*
sēgnis, -e, adj. *slow, sluggish*
sēmēsus, -a, -um, adj. *half-eaten*
sēmianimis, -e, adj. *half-dead*
sēminecis (no nom.), adj. *half-dead*
semper, adv. *always, ever*
senātus, -ūs, m. *senate, council of elders*
senecta, -ae, f. *old age*
senectūs, -ūtis, f. *old age*
senex, -is, m. *old man*
senior, -ōris, adj. *older, old*; m. *old man*
sequāx, -ācis, adj. *following*
sequor, 3, secūtus, dep. tr. *follow, pursue, aim at*
serēnus, -a, -um, adj. *clear, bright, fair* (of weather)

sermo, -ōnis, m. *speech*
sērus, -a, -um, adj. *late*
servo, 1, tr. *keep safe, preserve*
seu, see **sīve**
sīc, adv. *thus, in this way, so*
Sicānus, -a, -um, adj. *Sicanian*
Sīdŏnius, -a, -um, adj. *of or belonging to Sidon, Sidonian*
sīdus, -eris, n. *star*
signum, -i, n. *sign, signal*; pl. **signa**, -ōrum, *standards*
silva, -ae, f. *wood, forest*
Silvānus, -i, m. *god of woods and fields, Silvanus*
silvestris, -e, adj. *of the wood, forest-, wooded, woody*
similis, -e, adj. *like, alike* (sup. **simillimus**)
simul, adv. *at the same time*
sīn, conj. *but if*
singuli, -ae, -a, pl. adj. *one each, single, individual, one by one*
sino, 3, sīvi, situm, tr. *allow, let*
sisto, 3, stiti, statum, tr. *cause to stand, place, conduct*
sīve (**seu**), conj. *or if*; **seu...seu**, *either...or, whether...or*
socius, -i, m. *ally, comrade*
sodālis, -is, m. *comrade, fellow*
sōl, sōlis, m. *sun*
sōlācium, -ii, n. *comfort, solace*
sōlāmen, -inis, n. *consolation*
solium, -i, n. *seat, throne*
sollemnis, -e, adj. *yearly, established, stately, solemn*
solum, -i, n. *land, ground, soil*
sōlus, -a, -um, adj. *alone, only*
solvo, 3, solvi, solūtum, tr. *pay, loosen*
somnus, -i, m. *sleep, slumber*
sonitus, -ūs, m. *sound, crash, din*

VOCABULARY

sōpītus, -a, -um, p.p.p. as adj. *lulled to sleep, smouldering*
sopor, -ōris, m. *sleep, slumber*
soror, -ōris, f. *sister*
sors, -tis, f. *lot, share, destiny*
sōspes, -itis, adj. *safe*
spargo, 3, spārsi, spārsum, tr. *sprinkle, scatter*
spārsūrus, see **spargo**
spatium, -i, n. *space, space of time, interval, delay, respite*
spectātor, -ōris, m. *witness*
spectātus, -a, -um, p.p.p. as adj. *tried, proved*
specto, 1, tr. *watch, look at*
specula, -ae, f. *height, look-out*
spēlunca, -ae, f. *cave, cavern*
spēro, 1, tr. *hope, expect*
spēs, -ei, f. *hope, expectation*
spīna, -ae, f. *spine*
spīro, 1, intr. *breathe*
spolio, 1, tr. *despoil, deprive of*
spolium, -i, n. *spoil*
stagnum, -i, n. *pool*
sterno, 3, strāvi, strātum, tr. *lay low, throw down*; p.p.p. **strātus**, -a, -um, *lying*
Sthenius, -i, m. Latin warrior slain by Pallas
stirps, -is, f. *root, stock, stem*
sto, 1, steti, statum, intr. *stand, remain fixed*; **parvo stāre**, *cost little*
strāmen, -inis, n. *litter, straw* (anything which is strewn)
strātus, see **sterno**
strepitus, -ūs, m. *din, noise*
stringo, 3, strīnxi, strictum, tr. *graze, draw* (a sword); with **dē**, *scratch* or *wound slightly*
Strȳmonius, -a, -um, adj. *of* or *belonging to the Strymon* (a Thracian river), *Thracian*
stupeo, 2, -ui, intr. *be amazed*
Stygius, -a, -um, adj. *of Styx*, a river of Hades, or the underworld
suādeo, 2, suāsi, suāsum, intr. *advise, persuade* (with dat.)
sub, prep. with acc. *close to, towards*; with abl. *under, at, beneath, at foot of*
subeo, -īre, -ii, -itum, tr. and intr. with dat. *go towards, go to, take the place of, go up to*
subigo, 3, -ēgi, -āctum, tr. *incite, compel*
subitō, adv. *suddenly*
subitus, -a, -um, adj. *sudden*
subiungo, 3, -iūnxi, -iūnctum, tr. *bring under, subjugate*
sublāta, see **tollo**
subligo, 1, tr. *bind below, gird*
sublīmis, -e, adj. *above, on high*
subsisto, 3, -stiti, —, intr. *remain, halt, close up*
substiti, see **subsisto**
subter, prep. with acc. *beneath*
subveho, 3, -vēxi, -vectum, tr. *carry up, bring* or *carry upstream*
succēdo, 3, -cessi, -cessum, tr. and intr. *approach, enter, go below, follow up, succeed*
sūdum, -i, n. *clear weather*
sum, esse, fui, futūrus, intr. *be*
sūmo, 3, sūmpsi, sūmptum, tr. *take*; **poenās sūmere**, *exact the penalty*
super, adv. *moreover*; prep. with abl. *over*; with acc. *above, over, upon, beyond*

VOCABULARY

superbus, -a, -um, adj. *proud*
supero, 1, tr. and intr. *overcome*
superstes, -itis, adj. *surviving*
superus, -a, -um, adj. *above, upper*; pl. **superi**, *the gods above*
supplex, -icis, adj. *suppliant*
supplicium, -i, n. *punishment, penalty* (usually of death)
suprā, adv. and prep. with acc. *above*
suprēmus, -a, -um, superl. adj. *highest, top of, last*
sūrgo, 3, surrēxi, surrēctum, intr. *get up, arise, rise*
sūs, suis, c. *pig, sow*
suscito, 1, tr. *urge on, stir up*
suspicio, 3, -spexi, -spectum, tr. *look up at*
sustineo, 2, -tinui, -tentum, tr. *hold, withstand, sustain*
suus, -a, -um, poss. adj. *his, her, its, their own*

tābum (only nom., acc., and abl.), n. *corrupt moisture, matter*
tacitus, -a, -um, adj. *silent*
tālis, -e, adj. *such, of such a kind*
tamen, conj. *yet, still, after all*
tandem, adv. *at length, at last*
tango, 3, tetigi, tāctum, tr. *touch*
tantus, -a, -um, adj. *so great*
Tarcho, or **Tarchōn**, -ōnis, m. Etruscan leader
tardus, -a, -um, adj. *slow*
Tarpēius, -a, -um, adj. Roman proper name, *of the Tarpeian Rock* (on the edge of the Capitol in Rome)
Tartara, -ōrum, n. pl. *the lower world*
taurus, -i, m. *bull*
tēctum, -i, n., also pl., *roof, house, home*
Tegeaeus, -a, -um, adj. *of Tegea* (town in Arcadia)
tegmen, -inis, n. *covering*
tego, 3, tēxi, tēctum, tr. *hide*
tēla, -ae, f. *woven stuff, web, fabric*
tellūs, -ūris, f. *earth, land*
tēlum, -i, n. *weapon, spear*
temptāmentum, -i, n. *trial*
tempto, 1, tr. *try, essay, test*
tendo, 3, tetendi, tēnsum or tēntum, tr. *stretch, shoot, extend, direct*, intr. *go, make for, make one's way, strive*
tenebrae, -ārum, f. pl. *darkness*
teneo, 2, -ui, tentum, tr. *hold*
tenuis, -e, adj. *slender, thin, fine*
tepidus, -a, -um, adj. *warm*
ter, num. adv. *three times, thrice*
tergum, -i, n. *back, hide*; **dare tergum**, *flee, take to flight*; also pl. **terga**
terni, -ae, -a, distrib. num. adj. *three each, three*
terra, -ae, f. *earth, land*
terreo, 2, tr. *frighten, dismay*
terribilis, -e, adj. *terrible*
testor, 1, dep. tr. *bear witness, entreat, call to witness*
Teucri, -ōrum, m. pl. *Trojans*
Teuthrās, -ae, m. Trojan warrior
texo, 3, -xui, -xtum, tr. *weave*
thalamus, -i, m. *inner room*
Thoās, -antis, m. Trojan warrior slain by Halaesus
Thȳbris, -is or -idis, m. *Tiber*

VOCABULARY

Thymber, -bri, m. Latin warrior, twin-brother of Larides

Tiberīnus, -a, -um, adj. *of or belonging to the Tiber*

timor, -ōris, m. *fear, panic*

tolero, 1, tr. *support, sustain*

tollo, 3, sustuli, sublātum, tr. *take, take away, raise*

tono, 1, -ui, intr. *thunder, roar*

torrēns, -ntis, m. *torrent, stream*

torreo, 2, -ui, tostum, tr. *bake*

torus, -i, m. *mound, couch*

tosta, see **torreo**

totidem, indecl. num. adj. *just as many, of the same number*

tōtus, -a, -um, adj. *the whole*

traho, 3, trāxi, tractum, tr. *draw, drag, derive*

trāicio, 3, -iēci, -iectum, tr. *transfix, pierce, stab*

trānsverbero, 1, tr. *strike across, pierce*

tremendus, -a, -um, gerundive (tremo), *dread, fearful*

tremo, 3, -ui, tr. *tremble at*, intr. *tremble, quake*

trēs, trium, num. adj. *three*

trīginta, num. adj. *thirty*

trīstis, -e, adj. *gloomy, sad*

Trōia, -ae, f. Troy

Trōiānus, -a, -um, adj. *Trojan*

Trōiugena, -ae, c. *of Trojan birth or descent, Trojan-born*

Trōius, -a, -um, adj. *Trojan*

tropaeum, -i, n. *trophy*

truncus, -i, m. *tree-trunk*

trux, trucis, adj. *fierce, grim*

tū, tui, pron. *thou, you*

tuba, -ae, f. *trumpet*

tueor, 2, tuitus or tūtus, dep. tr. *see, gaze at*

tum, adv. *then, at that moment, at that time*

tumeo, 2, —, —, intr. *swell, be excited, be swollen*

tumidus, -a, -um, adj. *swollen*

tumor, -ōris, m. *swelling, wrath*

tumulus, -i, m. *mound, tomb*

tundo, 3, tutudi, tūnsum, tr. *beat*

tunica, -ae, f. *tunic*

turba, -ae, f. *crowd, throng*

turbo, 1, tr. *disturb, agitate*

Turnus, -i, m. Latin hero, leader of the Rutuli, and vassal prince of Latinus

tūs, tūris, n. *incense*

Tuscus, -a, -um, adj. *Etruscan*

tūtus, -a, -um, adj. *safe, secure*

tuus, -a, -um, poss. adj. *thy*

Typhōeus, -eos, m. *a giant slain by Hercules*

tyrannus, -i, m. *ruler, despot*

Tyrēs, acc. Tyrēn, Trojan warrior

Tyrrhēnus, -a, -um, adj. *Tyrrhenian, Etruscan*

ūber, -eris, n. *udder, teat*

ubi, adv. *when, where*

ūllus, -a, -um, adj. *any*

ulterius, comp. adv. *beyond*

ultrō, adv. *unasked, unprovoked*

umbra, -ae, f. *shade, darkness*

umbrōsus, -a, -um, adj. *full of shade, shady*

ūmecto, 1, tr. *bedew, wet*

umerus, -i, m. *shoulder*

ūnā, adv. *together, in company*

unda, -ae, f. *wave*

unde, adv. *whence, from where*

VOCABULARY

undique, adv. *on all sides*
unguis, -is, m. *nail, claw*
ungula, -ae, f. *hoof*
unguo, 3, ūnxi, ūnctum, tr. *anoint, oil, grease*
unquam, adv. *ever*
urbs, -is, f. *city*
ūnus, -a, -um, num. adj. *one*
urgeo, 2, ūrsi, tr. *force*
ursa, -ae, f. *she-bear*
ūsquam, adv. *anywhere*
ut, conj. with indic. *when, as*; with subj. *that, so that*
ūtor, 3, ūsus, dep. intr. *use, enjoy* (with abl.)

vaco, 1, intr. *be empty, be free*
vādo, 3, intr. *go*
vadum, -i, n. *ford, shallow water*; pl. *shallows, waters*
vāgīna, -ae, f. *sheath, scabbard*
valeo, 2, intr. *be strong, have strength*; **valē**, imperative, *farewell!, goodbye!*
validus, -a, -um, adj. *strong*
vānus, -a, -um, adj. *empty*
varius, -a, -um, adj. *various*
vāstus, -a, -um, adj. *huge, vast*
vātēs, -is, m. *seer, prophet*; f. *prophetess*
veho, 3, vēxi, vectum, tr. *bear, carry*; pass. *ride, drive, sail*
vello, 3, vulsi, vulsum, tr. *pluck, pluck out, pull up*
vēlo, 1, tr. *cover, deck, veil*
vēnātus, -ūs, m. *hunting, chase*
venio, 4, vēni, ventum, intr. *come*
Venus, -eris, f. *Venus (goddess of love), love*
verbum, -i, n. *word*

vērō, adv. *in truth, truly*
vertex, -icis, m. *head, summit*
verto, 3, verti, versum, tr. *turn, change, reverse, rout*
vērus, -a, -um, adj. *true, real*
vescor, 3, dep. intr. *eat, feed on* (with abl.)
Vesper, -eri and -eris, abl. -ere and -eri, m. *evening*
vestis, -is, f. *garment, clothes*
veto, 1, -ui, -itum, tr. *forbid*
vetus, -eris, adj. *old*
vetustus, -a, -um, adj. *ancient*
via, -ae, f. *way, track, path*
vibro, 1, tr. *shake, wield, brandish*, intr. *flash, quiver*
victor, -ōris, m. *conqueror, victor*; adj. *triumphant*
victus, -ūs, m. *victuals, food*
video, 2, vīdi, vīsum, tr. *see*; pass. *seem, seem best*
viduo, 1, tr. *deprive, bereave*
villōsus, -a, -um, adj. *hairy*
vīmen, -inis, n. *twig, switch, osier*
vincio, 4, vīnxi, vīnctum, tr. *bind, tie, fetter, fasten*
vinco, 3, vīci, victum, tr. *conquer, get beyond*
vinculum, -i, n. *bond, shackle*
vir, viri, m. *man, hero*
virga, -ae, f. *twig, sprout, branch, wand*
virgineus, -a, -um, adj. *of or belonging to a maiden*
viridis, -e, adj. *green*
virtūs, -ūtis, f. *valour, bravery*
vīs, vim, vī, f. *force*, pl. **vīrēs**, -ium, *strength*
viscera, -um, n. pl. *flesh, meat*
vīso, 3, vīsi, vīsum, tr. *visit, see*
vīsus, -ūs, m. *appearance, look*

VOCABULARY

vīta, -ae, f. *life, soul, existence*

vitta, -ae, f. *garland, fillet, ribbon*

vīvo, 3, vīxi, vīctum, intr. *be alive, live*

vix, adv. *scarcely, hardly*

voco, 1, tr. *call, summon*

Volcānius, -a, -um, adj. *of or belonging to Vulcan (the god of fire), fiery*

volgo, 1, tr. *spread abroad*

volo, 1, intr. *fly*

volo, velle, volui, irreg. tr. *wish, desire, be willing*

volŭcris, -is, f. *bird, winged or flying creature*

voluptās, -ātis, f. *delight*

volūtus, see **volvo**

volvo, 3, volvi, volūtum, tr. *roll*

vōtum, -i, n. *vow, prayer*

vōx, vōcis, f. *voice, cry, speech*; pl. *words, sounds*

vulnero, 1, tr. *wound, injure*

vulnus, -eris, n. *wound*

vultus, -ūs, m. *face, gaze, look, features, expression*; also pl.